Solutions Manual to Accompany Introductory Physics

Rebekah L. Mays and John D. Mays

Austin, Texas
2015

Second printing

Published by

novarescienceandmath.com

Printed in the United States of America

ISBN: 978-0-9966771-5-8

Novare Science & Math is an imprint of Novare Science & Math LLC.

For the complete catalog of textbooks and resources available from Novare Science & Math, visit novarescienceandmath.com.

Contents

Acknowledgement	iii
Preface	iv
Chapter 2	1
Chapter 3	8
Chapter 4	15
Chapter 5	27
Chapter 6	36
Chapter 7	43
Chapter 8	46
Chapter 9	56
Chapter 11	65
Chapter 13	74

NOVARE
SCIENCE & MATH

Acknowledgement

I (John) wish to express my gratitude to my daughter Rebekah Mays for carefully and meticulously compiling these solutions. Any errors that remain in this volume are my own responsibility.

Preface

This solutions manual contains fully detailed solutions for all of the computational problems contained in my text *Introductory Physics*. Teachers and students using that text should find this manual to be a valuable resource.

When comparing your results to the results shown here and to those in the text, keep in mind that the last digit is always uncertain because of the way significant digits in measurements are defined. When two results match except for a small difference in the most precise digit, we say that the results match. Because of rounding in calculators, it will not be uncommon for results shown here to differ from the answer key in the text or from your result by or two in the most precise digit.

I have checked and double checked the solutions to make them as accurate as possible. However, in any manual of this kind it is inevitable that errors remain. If you find an error, we would be much obliged if you would inform us of it by sending an email to info@novarescienceandmath.com.

Chapter 2 P. 45

Unit Conversions

1.

$$1{,}750\text{ m}\cdot\frac{100\text{ cm}}{1\text{ m}}\cdot\frac{1\text{ in}}{2.54\text{ cm}}\cdot\frac{1\text{ ft}}{12\text{ in}}=5{,}740\text{ ft}$$

2.

$$3.54\text{ g}\cdot\frac{1\text{ kg}}{1000\text{ g}}=0.00354\text{ kg}$$

3.

$$41.11\text{ mL}\cdot\frac{1\text{ L}}{1000\text{ mL}}=0.04111\text{ L}$$

4.

$$7\times10^{8}\text{ m}\cdot\frac{100\text{ cm}}{1\text{ m}}\cdot\frac{1\text{ in}}{2.54\text{ cm}}\cdot\frac{1\text{ ft}}{12\text{ in}}\cdot\frac{1\text{ mi}}{5{,}280\text{ ft}}=4\times10^{5}\text{ mi}$$

5.

$$1.5499\times10^{-12}\text{ mm}\cdot\frac{1\text{ m}}{1000\text{ mm}}=1.5499\times10^{-15}\text{ m}$$

6.

$$750\text{ cm}^3\cdot\frac{1\text{ mL}}{1\text{ cm}^3}\cdot\frac{1\text{ L}}{1000\text{ mL}}\cdot\frac{1\text{ m}^3}{1000\text{ L}}=7.5\times10^{-4}\text{ m}^3$$

7.

$$2.9979\times10^{8}\ \frac{\text{m}}{\text{s}}\cdot\frac{100\text{ cm}}{1\text{ m}}\cdot\frac{1\text{ in}}{2.54\text{ cm}}\cdot\frac{1\text{ ft}}{12\text{ in}}=9.836\times10^{8}\ \frac{\text{ft}}{\text{s}}$$

8.

$$168\text{ hr}\cdot\frac{60\text{ min}}{1\text{ hr}}\cdot\frac{60\text{ s}}{1\text{ min}}=605{,}000\text{ s}$$

9.

$$5{,}570\ \frac{\text{kg}}{\text{m}^3}\cdot\frac{1000\text{ g}}{1\text{ kg}}\cdot\frac{1\text{ m}^3}{1000\text{ L}}\cdot\frac{1\text{ L}}{1000\text{ mL}}\cdot\frac{1\text{ mL}}{1\text{ cm}^3}=5.57\ \frac{\text{g}}{\text{cm}^3}$$

10.

$$45\ \frac{\text{gal}}{\text{s}}\cdot\frac{3.786\ \text{L}}{1\ \text{gal}}\cdot\frac{1\ \text{m}^3}{1000\ \text{L}}\cdot\frac{60\ \text{s}}{1\ \text{min}}=1.0\times10^1\ \frac{\text{m}^3}{\text{min}}$$

11.

$$600{,}000\ \frac{\text{ft}^3}{\text{s}}\cdot\frac{(0.3048\ \text{m})^3}{1\ \text{ft}^3}\cdot\frac{1000\ \text{L}}{1\ \text{m}^3}\cdot\frac{60\ \text{s}}{1\ \text{min}}\cdot\frac{60\ \text{min}}{1\ \text{hr}}=6\times10^{10}\ \frac{\text{L}}{\text{hr}}$$

12.

$$5{,}200\ \text{mL}\cdot\frac{1\ \text{L}}{1000\ \text{mL}}\cdot\frac{1\ \text{m}^3}{1000\ \text{L}}=5.2\times10^{-3}\ \text{m}^3$$

13.

$$5.65\times10^2\ \text{mm}^2\cdot\frac{1\ \text{cm}}{10\ \text{mm}}\cdot\frac{1\ \text{cm}}{10\ \text{mm}}\cdot\frac{1\ \text{in}}{2.54\ \text{cm}}\cdot\frac{1\ \text{in}}{2.54\ \text{cm}}=0.876\ \text{in}^2$$

14.

$$32.16\ \frac{\text{ft}}{\text{s}^2}\cdot\frac{12\ \text{in}}{1\ \text{ft}}\cdot\frac{2.54\ \text{cm}}{1\ \text{in}}\cdot\frac{1\ \text{m}}{100\ \text{cm}}=9.802\ \frac{\text{m}}{\text{s}^2}$$

15.

$$5.001\ \frac{\mu\text{g}}{\text{s}}\cdot\frac{1\ \text{g}}{10^6\ \mu\text{g}}\cdot\frac{1\ \text{kg}}{1000\ \text{g}}\cdot\frac{60\ \text{s}}{1\ \text{min}}=3.001\times10^{-4}\ \frac{\text{kg}}{\text{min}}$$

16.

$$4.771\ \frac{\text{g}}{\text{mL}}\cdot\frac{1\ \text{kg}}{1000\ \text{g}}\cdot\frac{1000\ \text{mL}}{1\ \text{L}}\cdot\frac{1000\ \text{L}}{1\ \text{m}^3}=4{,}771\ \frac{\text{kg}}{\text{m}^3}$$

17.

$$13.6\ \frac{\text{g}}{\text{cm}^3}\cdot\frac{1000\ \text{mg}}{1\ \text{g}}\cdot\frac{100\ \text{cm}}{1\ \text{m}}\cdot\frac{100\ \text{cm}}{1\ \text{m}}\cdot\frac{100\ \text{cm}}{1\ \text{m}}=1.36\times10^{10}\ \frac{\text{mg}}{\text{m}^3}$$

18.

$$93{,}000{,}000\ \text{mi}\cdot\frac{5280\ \text{ft}}{1\ \text{mi}}\cdot\frac{0.3048\ \text{m}}{1\ \text{ft}}\cdot\frac{100\ \text{cm}}{1\ \text{m}}=1.5\times10^{13}\ \text{cm}$$

19.

$$65\ \frac{\text{mi}}{\text{hr}}\cdot\frac{5{,}280\ \text{ft}}{1\ \text{mi}}\cdot\frac{0.3048\ \text{m}}{1\ \text{ft}}\cdot\frac{1\ \text{hr}}{60\ \text{min}}\cdot\frac{1\ \text{min}}{60\ \text{s}}=29\ \frac{\text{m}}{\text{s}}$$

20.

$$633\text{ nm}\cdot\frac{1\text{ m}}{10^9\text{ nm}}\cdot\frac{100\text{ cm}}{1\text{ m}}\cdot\frac{1\text{ in}}{2.54\text{ cm}}=2.49\times10^{-5}\text{ in}$$

21.

$$0.05015\cdot 3.00\times10^8\ \frac{\text{m}}{\text{s}}\cdot\frac{60\text{ s}}{1\text{ min}}\cdot\frac{60\text{ min}}{1\text{ hr}}\cdot\frac{1\text{ ft}}{0.3048\text{ m}}\cdot\frac{1\text{ mi}}{5{,}280\text{ ft}}=3.37\times10^7\ \frac{\text{mi}}{\text{hr}}$$

Motion Exercises

p. 47

1.

$$d=25.1\text{ mi}\cdot\frac{5{,}280\text{ ft}}{1\text{ mi}}\cdot\frac{0.3048\text{ m}}{1\text{ ft}}=4.04\times10^4\text{ m}$$

$$t=0.50\text{ hr}\cdot\frac{60\text{ min}}{1\text{ hr}}\cdot\frac{60\text{ s}}{1\text{ min}}=1{,}800\text{ s}$$

$$v=?$$

$$d=vt$$

$$v=\frac{d}{t}=\frac{4.04\times10^4\text{ m}}{1{,}800\text{ s}}=22\ \frac{\text{m}}{\text{s}}$$

2.

$$22\ \frac{\text{m}}{\text{s}}\cdot\frac{1\text{ km}}{1000\text{ m}}\cdot\frac{60\text{ s}}{1\text{ min}}\cdot\frac{60\text{ min}}{1\text{ hr}}=79\ \frac{\text{km}}{\text{hr}}$$

3.

$$t=4.25\text{ hr}$$

$$v=5.0000\ \frac{\text{km}}{\text{hr}}$$

$$d=?$$

$$d=vt$$

$$d=5.0000\ \frac{\text{km}}{\text{hr}}\cdot 4.25\text{ hr}=21.3\text{ km}$$

4.

$$21.3\text{ km}\cdot\frac{1000\text{ m}}{1\text{ km}}\cdot\frac{1\text{ ft}}{0.3048\text{ m}}\cdot\frac{1\text{ mi}}{5{,}280\text{ ft}}=13.2\text{ mi}$$

5.

$$150.0\ \frac{\text{mi}}{\text{hr}} \cdot \frac{5{,}280\ \text{ft}}{1\ \text{mi}} \cdot \frac{0.3048\ \text{m}}{1\ \text{ft}} \cdot \frac{1\ \text{km}}{1000\ \text{m}} = 241.4\ \frac{\text{km}}{\text{hr}}$$

6.

$$v = 150.0\ \frac{\text{mi}}{\text{hr}} \cdot \frac{1\ \text{hr}}{60\ \text{min}} = 2.50\ \frac{\text{mi}}{\text{min}}$$

$$d = 10.0\ \text{mi}$$

$$t = ?$$

$$d = vt$$

$$t = \frac{d}{v}$$

$$t = \frac{10.0\ \text{mi}}{2.50\ \frac{\text{mi}}{\text{min}}} = 4.00\ \text{min}$$

7.

$$d = 3.0\ \text{km} \cdot \frac{1000\ \text{m}}{1\ \text{km}} = 3.0 \times 10^3\ \text{m}$$

$$t = 1\ \text{hr}\ 20.0\ \text{min} = 80.0\ \text{min} \cdot \frac{60\ \text{s}}{1\ \text{min}} = 4.80 \times 10^3\ \text{s}$$

$$v = ?$$

$$d = vt$$

$$v = \frac{d}{t} = \frac{3.0 \times 10^3\ \text{m}}{4.80 \times 10^3\ \text{s}} = 0.63\ \frac{\text{m}}{\text{s}}$$

8.

$$v_i = 0$$

$$v_f = 45\ \frac{\text{mi}}{\text{hr}} \cdot \frac{1\ \text{hr}}{60\ \text{min}} \cdot \frac{1\ \text{min}}{60\ \text{s}} \cdot \frac{5{,}280\ \text{ft}}{1\ \text{mi}} \cdot \frac{0.3048\ \text{m}}{1\ \text{ft}} = 20.1\ \frac{\text{m}}{\text{s}}$$

$$t = 36\ \text{s}$$

$$a = ?$$

$$a = \frac{v_f - v_i}{t} = \frac{20.1\frac{\text{m}}{\text{s}} - 0}{36\ \text{s}} = 0.56\ \frac{\text{m}}{\text{s}^2}$$

9.

$v_i = 31 \ \frac{\text{m}}{\text{s}}$

$t = 17 \text{ s}$

$v_f = 22 \ \frac{\text{m}}{\text{s}}$

$a = ?$

$$a = \frac{v_f - v_i}{t} = \frac{22 \ \frac{\text{m}}{\text{s}} - 31 \ \frac{\text{m}}{\text{s}}}{17 \text{ s}} = -0.53 \ \frac{\text{m}}{\text{s}^2}$$

10.

$d = 14.5 \text{ m}$

$v = c = 3.00 \times 10^8 \ \frac{\text{m}}{\text{s}}$

$t = ?$

$d = vt$

$$t = \frac{d}{v} = \frac{14.5 \text{ m}}{3.00 \times 10^8 \ \frac{\text{m}}{\text{s}}} = 4.83 \times 10^{-8} \text{ s} \cdot \frac{10^9 \text{ ns}}{\text{s}} 48.3 \text{ ns}$$

11.

$v_i = 0$

$v_f = 0.80 \cdot 3.00 \times 10^8 \ \frac{\text{m}}{\text{s}} = 2.40 \times 10^8 \ \frac{\text{m}}{\text{s}}$

$t = 18 \text{ hr } 6 \text{ min } 45 \text{ s} = 64800 \text{ s} + 360 \text{ s} + 45 \text{ s} = 65{,}205 \text{ s}$

$a = ?$

$$a = \frac{v_f - v_i}{t} = \frac{2.40 \times 10^8 \ \frac{\text{m}}{\text{s}} - 0}{65{,}205 \text{ s}} = 3{,}680 \ \frac{\text{m}}{\text{s}^2}$$

12.

$$d = 8.96\times10^{9}\ \text{km}\cdot\frac{1000\ \text{m}}{1\ \text{km}} = 8.96\times10^{12}\ \text{m}$$

$$v = 3.45\times10^{5}\ \frac{\text{m}}{\text{s}}$$

$$t = ?$$

$$d = vt$$

$$t = \frac{d}{v} = \frac{8.96\times10^{12}\ \text{m}}{3.45\times10^{5}\ \frac{\text{m}}{\text{s}}} = 2.597\times10^{7}\ \text{s}\cdot\frac{1\ \text{min}}{60\ \text{s}}\cdot\frac{1\ \text{hr}}{60\ \text{min}}\cdot\frac{1\ \text{day}}{24\ \text{hr}} = 301\ \text{days}$$

13.

$$a = 5.556\times10^{6}\ \frac{\text{cm}}{\text{s}^2}\cdot\frac{1\ \text{m}}{100\ \text{cm}} = 5.556\times10^{4}\ \frac{\text{m}}{\text{s}^2}$$

$$t = 45\ \text{ms}\cdot\frac{1\ \text{s}}{1000\ \text{ms}} = 4.5\times10^{-2}\ \text{s}$$

$$v_i = 0$$

$$v_f = ?$$

$$a = \frac{v_f - v_i}{t}$$

$$v_f = at + v_i = (5.556\times10^{4}\ \frac{\text{m}}{\text{s}^2})(4.5\times10^{-2}\ \text{s}) + (0\ \frac{\text{m}}{\text{s}}) = 2.5\times10^{3}\ \frac{\text{m}}{\text{s}}$$

14.

$$v_i = 4.005\times10^{3}\ \frac{\text{m}}{\text{s}}$$

$$a = 23.1\ \frac{\text{m}}{\text{s}^2}$$

$$t = 13.5\ \text{s}$$

$$v_f = ?$$

$$a = \frac{v_f - v_i}{t}$$

$$v_f = at + v_i = (23.1\ \frac{\text{m}}{\text{s}^2}\cdot 13.5\ \text{s}) + 4.005\times10^{3}\ \frac{\text{m}}{\text{s}} = 4.32\times10^{3}\ \frac{\text{m}}{\text{s}}$$

15.

$$v = c = 2.9979\times10^{8}\ \frac{\text{m}}{\text{s}}$$

$$d = 1.4965\times10^{8}\ \text{km}\cdot\frac{1000\ \text{m}}{1\ \text{km}} = 1.4965\times10^{11}\ \text{m}$$

$$t = ?$$

$$d = vt$$

$$t = \frac{d}{v} = \frac{1.4965\times10^{11}\ \text{m}}{2.9979\times10^{8}\ \frac{\text{m}}{\text{s}}} = 499.18\ \text{s}\cdot\frac{1\ \text{min}}{60\ \text{s}} = 8.3197\ \text{min}$$

Chapter 3 P.67

Newton's Second Law Practice Problems

1.

$m = 1{,}880 \text{ kg}$

$a = 1.50 \ \frac{\text{m}}{\text{s}^2}$

$F = ?$

$a = \frac{F}{m}$

$F = ma = 1{,}880 \text{ kg} \cdot 1.50 \ \frac{\text{m}}{\text{s}^2} = 2{,}820 \text{ N}$

2.

$m = 188.4 \text{ g} \cdot \frac{1 \text{ kg}}{1000 \text{ g}} = 0.1884 \text{ kg}$

$g = 9.80 \ \frac{\text{m}}{\text{s}^2}$

$F_w = ?$

$F_w = 0.1884 \text{ kg} \cdot 9.80 \ \frac{\text{m}}{\text{s}^2} = 1.85 \text{ N}$

3.

$F = 250.0 \text{ N}$

$m = 144{,}000 \text{ mg} \cdot \frac{1 \text{ g}}{1000 \text{ mg}} \cdot \frac{1 \text{ kg}}{1000 \text{ g}} = 0.144 \text{ kg}$

$a = ?$

$a = \frac{F}{m} = \frac{250.0 \text{ N}}{0.144 \text{ kg}} = 1{,}740 \ \frac{\text{m}}{\text{s}^2}$

4.

$$a = 2.3\ \frac{\text{m}}{\text{s}^2}$$

$$F = 230{,}000\ \text{N}$$

$$m = ?$$

$$a = \frac{F}{m}$$

$$m = \frac{F}{a} = \frac{230{,}000\ \text{N}}{2.3\ \frac{\text{m}}{\text{s}^2}} = 1.0 \times 10^5\ \text{kg}$$

5.

$$a = 0.0022\ \frac{\text{mi}}{\text{hr}^2} \cdot \frac{5{,}280\ \text{ft}}{1\ \text{mi}} \cdot \frac{0.3048\ \text{m}}{1\ \text{ft}} \cdot \frac{1\ \text{hr}}{60\ \text{min}} \cdot \frac{1\ \text{hr}}{60\ \text{min}} \cdot \frac{1\ \text{min}}{60\ \text{s}} \cdot \frac{1\ \text{min}}{60\ \text{s}} = 2.732 \times 10^{-7}\ \frac{\text{m}}{\text{s}^2}$$

$$m = 2.2\ \text{Mg} \cdot \frac{10^6\ \text{g}}{1\ \text{Mg}} \cdot \frac{1\ \text{kg}}{1000\ \text{g}} = 2.2 \times 10^3\ \text{kg}$$

$$F = ?$$

$$a = \frac{F}{m}$$

$$F = ma = 2.2 \times 10^3\ \text{kg} \cdot 2.732 \times 10^{-7}\ \frac{\text{m}}{\text{s}^2} = 6.0 \times 10^{-4}\ \text{N}$$

6.

$$F_w = 125.1\ \text{lb} \cdot \frac{4.45\ \text{N}}{1\ \text{lb}} = 556.7\ \text{N}$$

$$g = 9.80\ \frac{\text{m}}{\text{s}^2}$$

$$m = ?$$

$$F_w = mg$$

$$m = \frac{F_w}{g} = \frac{556.7\ \text{N}}{9.80\ \frac{\text{m}}{\text{s}^2}} = 56.8\ \text{kg}$$

7.

$$m = 56.8 \text{ kg}$$

$$F_w = 20.9 \text{ lb} \cdot \frac{4.45 \text{ N}}{1 \text{ lb}} = 93.01 \text{ N}$$

$$g_m = ?$$

$$F_w = mg_m$$

$$g_m = \frac{F_w}{m} = \frac{93.01 \text{ N}}{56.8 \text{ kg}} = 1.64 \ \frac{\text{m}}{\text{s}^2}$$

8.

$$v_i = 0$$

$$v_f = 125.0 \ \frac{\text{m}}{\text{s}}$$

$$t = 22.00 \text{ ms} \cdot \frac{1 \text{ s}}{1000 \text{ ms}} = 2.200 \times 10^{-2} \text{ s}$$

$$F = 142.0 \text{ N}$$

$$m = ?$$

$$a = \frac{v_f - v_i}{t} = \frac{125.0 \ \frac{\text{m}}{\text{s}} - 0}{2.200 \times 10^{-2} \text{ s}} = 5{,}681.8 \ \frac{\text{m}}{\text{s}^2}$$

$$a = \frac{F}{m}$$

$$m = \frac{F}{a} = \frac{142.0 \text{ N}}{5681.8 \ \frac{\text{m}}{\text{s}^2}} = 0.024992 \text{ kg} \cdot \frac{1000 \text{ g}}{1 \text{ kg}} = 24.99 \text{ g}$$

9.

$m = 4.5\text{ kg}$

$v_i = 0$

$$v_f = 8.00\ \frac{\text{mi}}{\text{hr}} \cdot \frac{1\text{ hr}}{60\text{ min}} \cdot \frac{1\text{ min}}{60\text{ s}} \cdot \frac{5{,}280\text{ ft}}{1\text{ mi}} \cdot \frac{0.3048\text{ m}}{1\text{ ft}} = 3.576\ \frac{\text{m}}{\text{s}}$$

$$t = 500\text{ ms} \cdot \frac{1\text{ s}}{1000\text{ ms}} = 0.5\text{ s}$$

$F = ?$

$$a = \frac{v_f - v_i}{t} = \frac{3.576\ \frac{\text{m}}{\text{s}} - 0}{0.5\text{ s}} = 7.152\ \frac{\text{m}}{\text{s}^2}$$

$$a = \frac{F}{m}$$

$$F = ma = 4.5\text{ kg} \cdot 7.152\ \frac{\text{m}}{\text{s}^2} = 30\text{ N}$$

10.

$$v_i = 2{,}500.0\ \frac{\text{km}}{\text{hr}} \cdot \frac{1\text{ hr}}{60\text{ min}} \cdot \frac{1\text{ min}}{60\text{ s}} \cdot \frac{1000\text{ m}}{1\text{ km}} = 694.4\ \frac{\text{m}}{\text{s}}$$

$t = 8.000\text{ s}$

$F = 45{,}450\text{ N}$

$$v_f = 2{,}750\ \frac{\text{km}}{\text{hr}} \cdot \frac{1\text{ hr}}{60\text{ min}} \cdot \frac{1\text{ min}}{60\text{ s}} \cdot \frac{1000\text{ m}}{1\text{ km}} = 763.9\ \frac{\text{m}}{\text{s}}$$

$m = ?$

$$a = \frac{v_f - v_i}{t} = \frac{763.89\ \frac{\text{m}}{\text{s}} - 694.44\ \frac{\text{m}}{\text{s}}}{8.000\text{ s}} = 8.681\ \frac{\text{m}}{\text{s}^2}$$

$$a = \frac{F}{m}$$

$$m = \frac{F}{a} = \frac{45{,}450\text{ N}}{8.681\ \frac{\text{m}}{\text{s}^2}} = 5{,}230\text{ kg}$$

11.

$$m = 166\text{ g} \cdot \frac{1\text{ kg}}{1000\text{ g}} = 0.166\text{ kg}$$

$$F = 0.0450\text{ N}$$

$$v_i = 0$$

$$t = 2.1\text{ s}$$

$$v_f = ?$$

$$a = \frac{F}{m} = \frac{0.0450\text{ N}}{0.166\text{ kg}} = 0.2711\ \frac{\text{m}}{\text{s}^2}$$

$$a = \frac{v_f - v_i}{t}$$

$$v_f = at + v_i = (0.2711\ \frac{\text{m}}{\text{s}^2} \cdot 2.1\text{ s}) + 0 = 0.57\ \frac{\text{m}}{\text{s}}$$

12.

$$m = 1.673 \times 10^{-18}\ \mu\text{g} \cdot \frac{1\text{ g}}{10^6\ \mu\text{g}} \cdot \frac{1\text{ kg}}{1000\text{ g}} = 1.673 \times 10^{-27}\text{ kg}$$

$$v_i = 0$$

$$v_f = c \cdot 0.0005 = 3.00 \times 10^8 \cdot 0.0005 = 1.50 \times 10^5\ \frac{\text{m}}{\text{s}}$$

$$t = 455\text{ ns} \cdot \frac{1\text{ s}}{10^9\text{ ns}} = 4.55 \times 10^{-7}\text{ s}$$

$$F = ?$$

$$a = \frac{v_f - v_i}{t} = \frac{1.50 \times 10^5\ \frac{\text{m}}{\text{s}} - 0}{4.55 \times 10^{-7}\text{ s}} = 3.30 \times 10^{11}\ \frac{\text{m}}{\text{s}^2}$$

$$a = \frac{F}{m}$$

$$F = ma = 1.673 \times 10^{-27}\text{ kg} \cdot 3.30 \times 10^{11}\ \frac{\text{m}}{\text{s}^2} = 5.52 \times 10^{-16}\text{ N} \cdot \frac{1\text{ GN}}{10^9\text{ N}} = 5.52 \times 10^{-25}\text{ GN}$$

13.

$$m = 6.548 \text{ Gg} \cdot \frac{10^9 \text{ g}}{1 \text{ Gg}} \cdot \frac{1 \text{ kg}}{1000 \text{ g}} = 6.548 \times 10^6 \text{ kg}$$

$$v_i = 8.35 \ \frac{\text{mi}}{\text{hr}} \cdot \frac{5{,}280 \text{ ft}}{1 \text{ mi}} \cdot \frac{0.3048 \text{ m}}{1 \text{ ft}} \cdot \frac{1 \text{ hr}}{60 \text{ min}} \cdot \frac{1 \text{ min}}{60 \text{ s}} = 3.732 \ \frac{\text{m}}{\text{s}}$$

$$v_f = 0$$

$$t = 0.288 \text{ min} \cdot \frac{60 \text{ s}}{1 \text{ min}} = 17.28 \text{ s}$$

$$F = ?$$

$$a = \frac{v_f - v_i}{t} = \frac{0 - 3.732 \ \frac{\text{m}}{\text{s}}}{17.28 \text{ s}} = -0.216 \ \frac{\text{m}}{\text{s}^2}$$

$$a = \frac{F}{m}$$

$$F = ma$$

$$F = 6.548 \times 10^6 \text{ kg} \cdot -0.216 \ \frac{\text{m}}{\text{s}^2} = 1.41 \times 10^6 \text{ N}$$

14.

$$v_i = 3.5 \ \frac{\text{cm}}{\text{s}} \cdot \frac{1 \text{ m}}{100 \text{ cm}} = 0.035 \ \frac{\text{m}}{\text{s}}$$

$$v_f = 18.5 \ \frac{\text{cm}}{\text{s}} \cdot \frac{1 \text{ m}}{100 \text{ cm}} = 0.185 \ \frac{\text{m}}{\text{s}}$$

$$t = 220 \text{ ms} \cdot \frac{1 \text{ s}}{1000 \text{ ms}} = 0.22 \text{ s}$$

$$a = ?$$

$$a = \frac{v_f - v_i}{t} = \frac{0.185 \ \frac{\text{m}}{\text{s}} - 0.035 \ \frac{\text{m}}{\text{s}}}{0.22 \text{ s}} = 0.68 \ \frac{\text{m}}{\text{s}^2}$$

15.

$$m = 45{,}500 \text{ kg}$$

$$v_i = 0 \ \frac{\text{m}}{\text{s}}$$

$$v_f = 55 \ \frac{\text{m}}{\text{s}}$$

$$t = 6.4 \text{ s}$$

$$a = ?$$

$$F = ?$$

$$a = \frac{v_f - v_i}{t} = \frac{55 \ \frac{\text{m}}{\text{s}} - 0}{6.4 \text{ s}} = 8.59 \ \frac{\text{m}}{\text{s}^2}$$

$$a = \boxed{8.6 \ \frac{\text{m}}{\text{s}^2}}$$

$$a = \frac{F}{m}$$

$$F = ma = 45{,}500 \text{ kg} \cdot 8.59 \ \frac{\text{m}}{\text{s}^2} = 3.9 \times 10^5 \text{ N}$$

16.

$$m = 8.5 \text{ g} \cdot \frac{1 \text{ kg}}{1000 \text{ g}} = 0.0085 \text{ kg}$$

$$a = 18{,}500 \ \frac{\text{m}}{\text{s}^2}$$

$$F = ?$$

$$a = \frac{F}{m}$$

$$F = ma = 0.0085 \text{ kg} \cdot 18{,}500 \ \frac{\text{m}}{\text{s}^2} = 160 \text{ N}$$

Chapter 4 p. 90

Classroom Energy Computation Examples

1.

$m = 1.00 \times 10^5$ kg

$h = 240 \text{ ft} \cdot \dfrac{0.3048 \text{ m}}{1 \text{ ft}} = 73.15 \text{ m}$

$g = 9.80 \ \dfrac{\text{m}}{\text{s}^2}$

$E_G = ?$

$E_G = mgh = 1.00 \times 10^5 \text{ kg} \cdot 9.80 \ \dfrac{\text{m}}{\text{s}^2} \cdot 73.15 \text{ m} = 72{,}000{,}000 \text{ J}$

2.

$m = 25 \text{ g} \cdot \dfrac{1 \text{ kg}}{1000 \text{ g}} = 0.025 \text{ kg}$

$v = 556 \ \dfrac{\text{ft}}{\text{s}} \cdot \dfrac{0.3048 \text{ m}}{1 \text{ ft}} = 169.5 \ \dfrac{\text{m}}{\text{s}}$

$E_K = ?$

$E_K = \dfrac{1}{2} m v^2 = \dfrac{1}{2} \cdot 0.025 \text{ kg} \cdot \left(169.5 \ \dfrac{\text{m}}{\text{s}} \right)^2 = 360 \text{ J}$

3.

$d = 75 \text{ cm} \cdot \dfrac{1 \text{ m}}{100 \text{ cm}} = 0.75 \text{ m}$

$m = 12{,}500 \text{ g} \cdot \dfrac{1 \text{ kg}}{1000 \text{ g}} = 12.5 \text{ kg}$

$W = ?$

$W = Fd$

$F_w = mg = 12.5 \text{ kg} \cdot 9.80 \ \dfrac{\text{m}}{\text{s}^2} = 122.5 \text{ kg}$

$W = 122.5 \text{ kg} \cdot 0.75 \text{ m} = 92 \text{ J}$

4.

$m = 12.5 \text{ kg}$

$h = 0.75 \text{ m}$

$$E_G = mgh = 0.75 \text{ m} \cdot 9.80 \ \frac{\text{m}}{\text{s}^2} \cdot 12.5 \text{ kg} = 92 \text{ J}$$

5.

$m = 12.5 \text{ kg}$

$h_i = 0.75 \text{ m}$

$h_f = 0$

$v_i = 0$

$v_f = ?$

$E_{Gi} + E_{Ki} = E_{Gf} + E_{Kf}$

$E_{Kf} = E_{Gi} + E_{Ki} - E_{Gf} = 92 \text{ J} + 0 - 0 = 92 \text{ J}$

$$v_f = \sqrt{\frac{2E_{Kf}}{m}} = \sqrt{\frac{2 \cdot 92 \text{ J}}{12.5 \text{ kg}}} = 3.8 \ \frac{\text{m}}{\text{s}}$$

6.

$$m = 255.8 \text{ g} \cdot \frac{1 \text{ kg}}{1000 \text{ g}} = 0.2558 \text{ kg}$$

$$h_i = 10.4 \text{ ft} \cdot \frac{0.3048 \text{ m}}{1 \text{ ft}} = 3.1699 \text{ m}$$

$$E_{Gi} = ?$$

$$E_{Gi} = mgh_i = 0.2558 \text{ kg} \cdot 9.80 \ \frac{\text{m}}{\text{s}^2} \cdot 3.1699 \text{ m} = 7.95 \text{ J}$$

$$h_f = 0$$

$$v_i = 0$$

$$v_f = ?$$

$$E_{Gi} + E_{Ki} = E_{Gf} + E_{Kf}$$

$$E_{Ki} = 0$$

$$E_{Gf} = 0$$

$$E_{Kf} = E_{Gi} + E_{Ki} - E_{Gf} = 7.95 \text{ J} + 0 \text{ J} - 0 \text{ J} = 7.95 \text{ J}$$

$$v_f = \sqrt{\frac{2E_{Kf}}{m}} = \sqrt{\frac{2 \cdot 7.95 \text{ J}}{0.2558 \text{ kg}}} = 7.88 \ \frac{\text{m}}{\text{s}}$$

Energy Calculations Set 1

1.

$$m = 1.31 \times 10^3 \text{ kg}$$

$$h = 177.44 \text{ ft} \cdot \frac{0.3048 \text{ m}}{1 \text{ ft}} = 54.084 \text{ m}$$

$$E_G = ?$$

$$E_G = mgh = 1.31 \times 10^3 \text{ kg} \cdot 9.80 \ \frac{\text{m}}{\text{s}^2} \cdot 54.084 \text{ m} = 694{,}000 \text{ J}$$

2.

$$m = 2{,}345 \text{ kg}$$

$$v = 31 \ \frac{\text{mi}}{\text{hr}} \cdot \frac{5{,}280 \text{ ft}}{1 \text{ mi}} \cdot \frac{0.3048 \text{ m}}{1 \text{ ft}} \cdot \frac{1 \text{ hr}}{60 \text{ min}} \cdot \frac{1 \text{ min}}{60 \text{ s}} = 13.858 \ \frac{\text{m}}{\text{s}}$$

$$E_K = \frac{1}{2} mv^2 = \frac{1}{2} \cdot 2{,}345 \text{ kg} \cdot (13.858 \ \frac{\text{m}}{\text{s}})^2 = 230{,}000 \text{ J}$$

3.

$$d = 61.7 \text{ cm} \cdot \frac{1 \text{ m}}{100 \text{ cm}} = 0.617 \text{ m}$$

$$m = 17.5 \text{ kg}$$

$$W = ?$$

$$a = \frac{F}{m}$$

$$F_w = mg = 17.5 \text{ kg} \cdot 9.80 \ \frac{\text{m}}{\text{s}^2} = 171.5 \text{ N}$$

$$W = F_w d = 171.5 \text{ N} \cdot 0.617 \text{ m} = 106 \text{ J}$$

4.

$$h = 61.7 \text{ cm} \cdot \frac{1 \text{ m}}{100 \text{ cm}} = 0.617 \text{ m}$$

$$m = 17.5 \text{ kg}$$

$$E_G = ?$$

$$E_G = mgh = 17.5 \text{ kg} \cdot 9.80 \ \frac{\text{m}}{\text{s}^2} \cdot 0.617 \text{ m} = 106 \text{ J}$$

5.

$$m = 17.5 \text{ kg}$$

$$h_f = 0.617 \text{ m}$$

$$h_i = 0$$

$$v_i = 0$$

$$v_f = ?$$

$$E_{Gi} + E_{Ki} = E_{Gf} + E_{Kf}$$

$$E_{Kf} = E_{Gi} + E_{Ki} - E_{Gf} = 106 \text{ J} + 0 \text{ J} - 0 \text{ J} = 106 \text{ J}$$

$$v_f = \sqrt{\frac{2E_{Kf}}{m}} = \sqrt{\frac{2 \cdot 106 \text{ J}}{17.5 \text{ kg}}} = 12.1 \ \frac{\text{m}}{\text{s}} = 3.48 \ \frac{\text{m}}{\text{s}}$$

6.

$$m = 122\ \text{g} \cdot \frac{1\ \text{kg}}{1000\ \text{g}} = 0.122\ \text{kg}$$

$$v_i = 13.75\ \frac{\text{m}}{\text{s}}$$

$$v_f = 0$$

$$h_i = 0$$

$$h_f = ?$$

$$E_{Ki} = \frac{1}{2}mv^2 = \frac{1}{2} \cdot 0.122\ \text{kg} \cdot (13.75\ \frac{\text{m}}{\text{s}})^2 = 11.53\ \text{J}$$

$$E_{Gi} + E_{Ki} = E_{Gf} + E_{Kf}$$

$$E_{Gf} = E_{Gi} + E_{Ki} - E_{Kf} = 0 + 11.53\ \text{J} - 0 = 11.53\ \text{J}$$

$$E_{Gf} = mgh_f$$

$$h_f = \frac{E_{Gf}}{mg} = \frac{11.53\ \text{J}}{0.122\ \text{kg} \cdot 9.8\ \frac{\text{m}}{\text{s}^2}} = 9.65\ \text{m}$$

7.

$$m = 325\ \text{g} \cdot \frac{1\ \text{kg}}{1000\ \text{g}} = 0.325\ \text{kg}$$

$$h_i = 36.1\ \text{m}$$

$$h_f = 0$$

$$v_i = 0$$

$$v_f = ?$$

$$E_{Gi} = mgh_i = 0.325\ \text{kg} \cdot 9.80\ \frac{\text{m}}{\text{s}^2} \cdot 36.1\ \text{m} = 114.98\ \text{J}$$

$$E_{Gi} + E_{Ki} = E_{Gf} + E_{Kf}$$

$$E_{Kf} = E_{Gi} + E_{Ki} - E_{Gf} = 114.98\ \text{J} + 0 - 0$$

$$E_{Kf} = 114.98\ \text{J}$$

$$E_{Kf} = \frac{1}{2}mv^2$$

$$v_f = \sqrt{\frac{2E_{Kf}}{m}} = \sqrt{\frac{2 \cdot 114.98\ \text{J}}{0.325\ \text{kg}}} = 26.6\ \frac{\text{m}}{\text{s}}$$

8.

$F = 735 \text{ N}$

$d = 26 \text{ m}$

$W = ?$

$W = Fd = 735 \text{ N} \cdot 26 \text{ m} = 19{,}000 \text{ J}$

Energy Calculations Set 2

1.a.

$$F_w = 20 \cdot 80.0 \text{ lb} \cdot \frac{4.45 \text{ N}}{1 \text{ lb}} = 7{,}120 \text{ N}$$

$h = 8.5 \text{ m}$

$$g = 9.80 \ \frac{\text{m}}{\text{s}^2}$$

$m = ?$

$F_w = mg$

$$m = \frac{F_w}{g}$$

$$m = \frac{7{,}120 \text{ N}}{9.80 \ \frac{\text{m}}{\text{s}^2}} = 727 \text{ kg}$$

1.b

$F = 7{,}120 \text{ N}$

$d = 8.5 \text{ m}$

$W = Fd = 7{,}120 \text{ N} \cdot 8.5 \text{ m}$

$W = 6.0 \times 10^4 \text{ J}$

1.c.

$m = 727 \text{ kg}$

$$g = 9.80 \ \frac{\text{m}}{\text{s}^2}$$

$h = 8.5 \text{ m}$

$E_G = mgh$

$$E_G = 727 \text{ kg} \cdot 9.80 \ \frac{\text{m}}{\text{s}^2} \cdot 8.5 \text{ m} = 6.0 \times 10^4 \text{ J}$$

1.d.

$m = 727 \text{ kg}$

$v_i = 0$

$E_{Ki} = ?$

$E_{Ki} = \frac{1}{2} m v_i^{\ 2} = \frac{1}{2} \cdot 727 \text{ kg} \cdot 0^2 = 0 \text{ J}$

$E_{Gi} = 6.0 \times 10^4 \text{ J}$

$E_{Gf} = 0$

$E_{Kf} = ?$

$E_{Gi} + E_{Ki} = E_{Gf} + E_{Kf}$

$E_{Kf} = E_{Gi} + E_{Ki} - E_{Gf} = 6.0 \times 10^4 \text{ J} + 0 - 0 = 6.0 \times 10^4 \text{ J}$

1.e.

$m = 727 \text{ kg}$

$E_{Kf} = 6.0 \times 10^4 \text{ J}$

$v_f = ?$

$E_{Kf} = \frac{1}{2} m v_f^{\ 2}$

$v_f = \sqrt{\frac{2E_{Kf}}{m}} = \sqrt{\frac{2 \cdot 6.0 \times 10^4 \text{ J}}{727 \text{ kg}}} = 13 \ \frac{\text{m}}{\text{s}}$

2.a.

$$F_W = 3{,}193 \text{ lb} \cdot \frac{4.45 \text{ N}}{1 \text{ lb}} = 14{,}209 \text{ N}$$

$$m = ?$$

$$F_W = mg$$

$$m = \frac{F_W}{g}$$

$$m = \frac{14{,}209 \text{ N}}{9.80 \ \frac{\text{m}}{\text{s}^2}} = 1{,}450 \text{ kg}$$

$$h = 16 \text{ m}$$

$$E_G = mgh = 1{,}450 \text{ kg} \cdot 9.80 \ \frac{\text{m}}{\text{s}^2} \cdot 16 \text{ m} = 227{,}360 \text{ J}$$

$$E_G = 230{,}000 \text{ J}$$

2.c.

$$E_{Kf} = ?$$

$$E_{G_i} + E_{Ki} = E_{Gf} + E_{Kf}$$

$$E_{Kf} = E_{G_i} + E_{Ki} - E_{Gf}$$

$$E_{Kf} = 230{,}000 \text{ J} + 0 - 0 = 230{,}000 \text{ J}$$

2.d.

$$E_{Kf} = 230{,}000 \text{ J}$$

$$m = 1{,}450 \text{ kg}$$

$$v_f = ?$$

$$v_f = \sqrt{\frac{2E_{Kf}}{m}} = \sqrt{\frac{2 \cdot 230{,}000 \text{ J}}{1{,}450 \text{ kg}}} = 18 \ \frac{\text{m}}{\text{s}}$$

Energy Calculations Set 3

1.a.

$$F_w = 27.05\ \text{lb} \cdot \frac{4.45\ \text{N}}{1\ \text{lb}} = 120.37\ \text{N}$$

$$d = 185\ \text{cm} \cdot \frac{1\ \text{m}}{100\ \text{cm}} = 1.85\ \text{m}$$

$$W = Fd = 120.37\ \text{N} \cdot 1.85\ \text{m} = 223\ \text{J}$$

1.b

$$F_w = 120.37\ \text{N}$$

$$h = 1.85\ \text{m}$$

$$g = 9.80\ \frac{\text{m}}{\text{s}^2}$$

$$m = ?$$

$$F_w = mg$$

$$m = \frac{F_w}{g}$$

$$m = \frac{120.37\ \text{N}}{9.80\ \frac{\text{m}}{\text{s}^2}} = 12.28\ \text{kg}$$

$$E_G = ?$$

$$E_G = mgh = 12.28\ \text{kg} \cdot 9.80\ \frac{\text{m}}{\text{s}^2} \cdot 1.85\ \text{m} = 223\ \text{J}$$

1.c.

$$m = 12.28\ \text{kg}$$

$$g = 9.80\ \frac{\text{m}}{\text{s}^2}$$

$$h = 0\ \text{m}$$

$$E_G = ?$$

$$E_G = mgh = 12.28\ \text{kg} \cdot 9.80\ \frac{\text{m}}{\text{s}^2} \cdot 0\ \text{m} = 0\ \text{J}$$

1.d.

$E_{G_i} = 222.7 \text{ J}$

$E_{G_f} = 0 \text{ J}$

$E_{Ki} = 0 \text{ J}$

$E_{Kf} = ?$

$E_{Gi} + E_{Ki} = E_{G_f} + E_{Kf}$

$E_{Kf} = E_{Gi} + E_{Ki} - E_{G_f} = 222.7 \text{ J} + 0 - 0 = 223 \text{ J}$

1.e.

$E_{Kf} = 222.7 \text{ J}$

$m = 12.28 \text{ kg}$

$v_f = ?$

$$v_f = \sqrt{\frac{2E_{Kf}}{m}} = \sqrt{\frac{2 \cdot 222.7 \text{ J}}{12.28 \text{ kg}}} = 6.02 \ \frac{\text{m}}{\text{s}}$$

2.

$$d = 197 \text{ ft} \cdot \frac{0.3048 \text{ m}}{1 \text{ ft}} = 60.05 \text{ m}$$

$m = 6.016 \times 10^6 \text{ kg}$

$$F_w = mg = 6.016 \times 10^6 \text{ kg} \cdot 9.80 \ \frac{\text{m}}{\text{s}^2} = 5.896 \times 10^7 \text{ N}$$

$W = ?$

$W = F_w d = 5.896 \times 10^7 \text{ N} \cdot 60.05 \text{ m}$

$W = 3.54 \times 10^9 \text{ J}$

3.

$m = 5{,}122 \text{ kg}$

$h_i = 25.0 \text{ m}$

$v_f = ?$

$E_{Gi} = mgh_i = 5{,}122 \text{ kg} \cdot 9.80 \ \frac{\text{m}}{\text{s}^2} \cdot 25.0 \text{ m} = 1.2549 \times 10^6 \text{ J}$

$E_{Gi} = E_{Kf} = 1.2549 \times 10^6 \text{ J}$

$v_f = \sqrt{\frac{2E_{Kf}}{m}} = \sqrt{\frac{2 \cdot 1.2549 \times 10^6 \text{ J}}{5{,}122 \text{ kg}}} = 22.1 \ \frac{\text{m}}{\text{s}}$

4.a.

$F_w = 104.6 \text{ lb} \cdot \frac{4.45 \text{ N}}{1 \text{ lb}} = 4.6547 \times 10^2 \text{ N}$

$d = 13 \text{ steps} \cdot \frac{16.5 \text{ cm}}{1 \text{ step}} \cdot \frac{1 \text{ m}}{100 \text{ cm}} = 2.145 \text{ m}$

$W = ?$

$W = F_w d = 4.6547 \times 10^2 \text{ N} \cdot 2.145 \text{ m} = 998 \text{ J}$

4.b.

$F_w = 4.6547 \times 10^2 \text{ N}$

$F_w = mg$

$m = \frac{F_w}{g} = \frac{4.6547 \times 10^2 \text{ N}}{9.80 \ \frac{\text{m}}{\text{s}^2}} = 47.497 \text{ kg}$

$h_i = 2.145 \text{ m}$

$v_f = ?$

$E_{Gi} = mgh_i = 47.497 \text{ kg} \cdot 9.80 \ \frac{\text{m}}{\text{s}^2} \cdot 2.145 \text{ m}$

$E_{Gi} = 998.43 \text{ J}$

$E_{Gi} + E_{Ki} = E_{Gf} + E_{Kf}$

$E_{Kf} = E_{Gi} + E_{Ki} - E_{Gf} = 998.43 \text{ J} + 0 - 0$

$E_{Kf} = 998.43 \text{ J}$

$v_f = \sqrt{\frac{2E_{Kf}}{m}} = \sqrt{\frac{2 \cdot 998.43 \text{ J}}{47.497 \text{ kg}}} = 6.48 \ \frac{\text{m}}{\text{s}}$

5.

$$m = 351 \text{ g} \cdot \frac{1 \text{ kg}}{1000 \text{ g}} = 0.351 \text{ kg}$$

$$v_i = 500.00 \ \frac{\text{cm}}{\text{s}} \cdot \frac{1 \text{ m}}{100 \text{ cm}} = 5.00 \ \frac{\text{m}}{\text{s}}$$

$$v_f = 0 \ \frac{\text{m}}{\text{s}}$$

$$h_i = 0 \text{ m}$$

$$h_f = ?$$

$$E_{Ki} = \frac{1}{2} m v^2 = \frac{1}{2} \cdot 0.351 \text{ kg} \cdot (5.00 \ \frac{\text{m}}{\text{s}})^2$$

$$E_{Ki} = 4.388 \text{ J}$$

$$E_{Gi} + E_{Ki} = E_{Gf} + E_{Kf}$$

$$E_{Gf} = E_{Gi} + E_{Ki} - E_{Kf} = 0 \text{ J} + 4.388 \text{ J} - 0 \text{ J}$$

$$E_{Gf} = 4.388 \text{ J}$$

$$E_{Gf} = m g h_f$$

$$h_f = \frac{E_{Gf}}{mg} = \frac{4.388 \text{ J}}{0.351 \text{ kg} \cdot 9.80 \ \frac{\text{m}}{\text{s}^2}} = 1.28 \text{ m}$$

Chapter 5

Basic Momentum Computations

1.

$m = 1350 \text{ kg}$

$v = 35 \dfrac{\text{mi}}{\text{hr}} \cdot \dfrac{5{,}280 \text{ ft}}{\text{mi}} \cdot \dfrac{0.3048 \text{ m}}{1 \text{ ft}} \cdot \dfrac{1 \text{ hr}}{3{,}600 \text{ s}} = 15.6 \dfrac{\text{m}}{\text{s}}$

$p = ?$

$p = mv = 1350 \text{ kg} \cdot 15.6 \dfrac{\text{m}}{\text{s}} = 21{,}000 \dfrac{\text{kg} \cdot \text{m}}{\text{s}}$

2.

$m = 8.71 \times 10^{-4} \text{ g} \cdot \dfrac{1 \text{ kg}}{1000 \text{ g}} = 8.71 \times 10^{-7} \text{ kg}$

$v = 725.5 \dfrac{\text{m}}{\text{s}}$

$p = ?$

$p = mv = 8.71 \times 10^{-7} \text{ kg} \cdot 725.5 \dfrac{\text{m}}{\text{s}} = 6.32 \times 10^{-4} \dfrac{\text{kg} \cdot \text{m}}{\text{s}}$

3.

$m = 144.3 \text{ g} \cdot \dfrac{1 \text{ kg}}{1000 \text{ g}} = 0.1443 \text{ kg}$

$v = 99.55 \dfrac{\text{mi}}{\text{hr}} \cdot \dfrac{5{,}280 \text{ ft}}{1 \text{ mi}} \cdot \dfrac{0.3048 \text{ m}}{1 \text{ ft}} \cdot \dfrac{1 \text{ hr}}{3{,}600 \text{ s}} = 44.50 \dfrac{\text{m}}{\text{s}}$

$p = ?$

$p = mv = 0.1443 \text{ kg} \cdot 44.503 \dfrac{\text{m}}{\text{s}} = 6.421 \dfrac{\text{kg} \cdot \text{m}}{\text{s}}$

4.

$$v = 1{,}847\ \frac{\text{ft}}{\text{s}} \cdot \frac{0.3048\ \text{m}}{1\ \text{ft}} = 5.630\times10^{2}\ \frac{\text{m}}{\text{s}}$$

$$p = 25{,}565\ \frac{\text{kg}\cdot\text{m}}{\text{s}}$$

$$m = ?$$

$$F_w = ?$$

$$p = mv$$

$$m = \frac{p}{v} = \frac{25{,}565\ \frac{\text{kg}\cdot\text{m}}{\text{s}}}{5.630\times10^{2}\ \frac{\text{m}}{\text{s}}} = 45.41\ \text{kg}$$

$$F_w = mg = 45.41\ \text{kg}\cdot 9.80\ \frac{\text{m}}{\text{s}^2} = 445.02\ \text{N}\cdot\frac{1\ \text{lb}}{4.45\ \text{N}} = 1.00\times10^{2}\ \text{lb}$$

5.

$$m = 9.11\times10^{-28}\ \text{g}\cdot\frac{1\ \text{kg}}{1000\ \text{g}} = 9.11\times10^{-31}\ \text{kg}$$

$$v = 0.3375\cdot 3.00\times10^{8}\ \frac{\text{m}}{\text{s}} = 1.0125\times10^{8}\ \frac{\text{m}}{\text{s}}$$

$$p = ?$$

$$p = mv = 9.11\times10^{-31}\ \text{kg}\cdot 1.0125\times10^{8}\ \frac{\text{m}}{\text{s}} = 9.22\times10^{-23}\ \frac{\text{kg}\cdot\text{m}}{\text{s}}$$

6.

$$m = 1.673\times10^{-24}\ \text{g}\cdot\frac{1\ \text{kg}}{1000\ \text{g}} = 1.673\times10^{-27}\ \text{kg}$$

$$p = 4.5516\times10^{-23}\ \frac{\text{kg}\cdot\text{m}}{\text{s}}$$

$$v = ?$$

$$p = mv$$

$$v = \frac{p}{m} = \frac{4.5516\times10^{-23}\ \frac{\text{kg}\cdot\text{m}}{\text{s}}}{1.673\times10^{-27}\ \text{kg}} = 2.721\times10^{4}\ \frac{\text{m}}{\text{s}}$$

$$2.721\times10^{4}\ \frac{\text{m}}{\text{s}}\cdot\frac{100\ \text{cm}}{1\ \text{m}}\cdot\frac{1\ \text{in}}{2.54\ \text{cm}} = 1.071\times10^{6}\ \frac{\text{in}}{\text{s}}$$

7.

$$F_w = 47{,}055 \text{ tons} \cdot \frac{2{,}000 \text{ lb}}{1 \text{ ton}} \cdot \frac{4.45 \text{ N}}{1 \text{ lb}} = 4.1879 \times 10^8 \text{ N}$$

$$v = 55.75 \ \frac{\text{mi}}{\text{hr}} \cdot \frac{5{,}280 \text{ ft}}{1 \text{ mi}} \cdot \frac{0.3048 \text{ m}}{1 \text{ ft}} \cdot \frac{1 \text{ hr}}{3{,}600 \text{ s}} = 24.92 \ \frac{\text{m}}{\text{s}}$$

$$p = ?$$

$$F_w = mg$$

$$m = \frac{F_w}{g} = \frac{4.1879 \times 10^8 \text{ N}}{9.80 \ \frac{\text{m}}{\text{s}^2}} = 4.27 \times 10^7 \text{ kg}$$

$$p = mv = 4.27 \times 10^7 \text{ kg} \cdot 24.92 \ \frac{\text{m}}{\text{s}} = 1{,}060{,}000{,}000 \ \frac{\text{kg} \cdot \text{m}}{\text{s}}$$

8.

$$m = 5.98 \times 10^{24} \text{ kg}$$

$$l = 5.84 \times 10^8 \text{ mi} \cdot \frac{5{,}280 \text{ ft}}{1 \text{ mi}} \cdot \frac{0.3048 \text{ m}}{1 \text{ ft}} = 9.399 \times 10^{11} \text{ m}$$

$$p = ?$$

$$t = \frac{365 \text{ days}}{1 \text{ orbit}} \cdot \frac{24 \text{ hr}}{1 \text{ day}} \cdot \frac{3{,}600 \text{ s}}{1 \text{ hr}} = 3.15 \times 10^7 \text{ s}$$

$$v = \frac{9.399 \times 10^{11} \text{ m}}{3.15 \times 10^7 \text{ s}} = 2.98 \times 10^4 \ \frac{\text{m}}{\text{s}}$$

$$p = mv = 5.98 \times 10^{24} \text{ kg} \cdot 2.98 \times 10^4 \ \frac{\text{m}}{\text{s}} = 1.78 \times 10^{29} \ \frac{\text{kg} \cdot \text{m}}{\text{s}}$$

9.

$$v = 13.25 \ \frac{\text{in}}{\text{s}} \cdot \frac{2.54 \text{ cm}}{1 \text{ in}} \cdot \frac{1 \text{ m}}{100 \text{ cm}} = 0.3366 \ \frac{\text{m}}{\text{s}}$$

$$p = 5.9805 \times 10^{-4} \ \frac{\text{kg} \cdot \text{m}}{\text{s}}$$

$$m = ?$$

$$p = mv$$

$$m = \frac{p}{v} = \frac{5.9805 \times 10^{-4} \ \frac{\text{kg} \cdot \text{m}}{\text{s}}}{0.3366 \ \frac{\text{m}}{\text{s}}} = 0.001777 \text{ kg} \cdot \frac{1000 \text{ g}}{1 \text{ kg}} = 1.777 \text{ g}$$

10.

$$p = 2{,}177.9\ \frac{\text{kg}\cdot\text{m}}{\text{s}}$$

$$F_w = 188.5\ \text{lb}\cdot\frac{4.45\ \text{N}}{1\ \text{lb}} = 838.83\ \text{N}$$

$$v = ?$$

$$F_w = mg$$

$$m = \frac{F_w}{g} = \frac{838.83\ \text{N}}{9.80\ \frac{\text{m}}{\text{s}^2}} = 85.594\ \text{kg}$$

$$p = mv$$

$$v = \frac{p}{m} = \frac{2{,}177.9\ \frac{\text{kg}\cdot\text{m}}{\text{s}}}{85.594\ \text{kg}} = 25.4\ \frac{\text{m}}{\text{s}}$$

Conservation of Momentum Computations

2.

$$v_{1i} = 2.00\ \frac{\text{mi}}{\text{hr}}\cdot\frac{5{,}280\ \text{ft}}{1\ \text{mi}}\cdot\frac{0.3048\ \text{m}}{1\ \text{ft}}\cdot\frac{1\ \text{hr}}{60\ \text{min}}\cdot\frac{1\ \text{min}}{60\ \text{s}} = 0.894\ \frac{\text{m}}{\text{s}}$$

$$v_{1f} = 1.00\ \frac{\text{mi}}{\text{hr}}\cdot\frac{5{,}280\ \text{ft}}{1\ \text{mi}}\cdot\frac{0.3048\ \text{m}}{1\ \text{ft}}\cdot\frac{1\ \text{hr}}{60\ \text{min}}\cdot\frac{1\ \text{min}}{60\ \text{s}} = 0.447\ \frac{\text{m}}{\text{s}}$$

$$m_1 = 3{,}450\ \text{kg}$$

$$m_2 = 1{,}150\ \text{kg}$$

$$v_{2f} = ?$$

$$p_{1i} = p_{1f} + p_{2f}$$

$$p_{2f} = p_{1i} - p_{1f}$$

$$p_{1i} = m_1 v_{1i} = 3{,}450\ \text{kg}\cdot 0.894\ \frac{\text{m}}{\text{s}} = 3{,}084\ \frac{\text{kg}\cdot\text{m}}{\text{s}}$$

$$p_{1f} = m_1 v_{1f} = 3{,}450\ \text{kg}\cdot 0.447\ \frac{\text{m}}{\text{s}} = 1{,}542\ \frac{\text{kg}\cdot\text{m}}{\text{s}}$$

$$p_{2f} = 3{,}084\ \frac{\text{kg}\cdot\text{m}}{\text{s}} - 1{,}542\ \frac{\text{kg}\cdot\text{m}}{\text{s}} = 1{,}542\ \frac{\text{kg}\cdot\text{m}}{\text{s}}$$

$$v_{2f} = \frac{p_{2f}}{m_2} = \frac{1{,}542\ \frac{\text{kg}\cdot\text{m}}{\text{s}}}{1{,}150\ \text{kg}} = 1.34\ \frac{\text{m}}{\text{s}}\cdot\frac{3{,}600\ \text{s}}{1\ \text{hr}}\cdot\frac{1\ \text{ft}}{0.3048\ \text{m}}\cdot\frac{1\ \text{mi}}{5{,}280\ \text{ft}} = 3.00\ \frac{\text{mi}}{\text{hr}}$$

3.

$m_1 = 3{,}450\ \text{kg}$

$m_2 = 1{,}150\ \text{kg}$

$v_{1i} = 0.89408\ \dfrac{\text{m}}{\text{s}}$

$v_{2i} = 0\ \dfrac{\text{m}}{\text{s}}$

$v_{1f} = 0.447\ \dfrac{\text{m}}{\text{s}}$

$v_{2f} = 1.34\ \dfrac{\text{m}}{\text{s}}$

$E_{Ki} = ?$

$E_{Kf} = ?$

$$E_{Ki} = \frac{1}{2} m_1 {v_{1i}}^2 = \frac{1}{2} \cdot 3{,}450\ \text{kg} \cdot (0.894\ \frac{\text{m}}{\text{s}})^2 = 1{,}380\ \text{J}$$

$$E_{Kf} = \frac{1}{2} m_1 {v_{1f}}^2 + \frac{1}{2} m_2 {v_{2f}}^2 = \left(\frac{1}{2} \cdot 3{,}450\ \text{kg} \cdot \left(0.447\ \frac{\text{m}}{\text{s}} \right)^2 \right) + \left(\frac{1}{2} \cdot 1{,}150\ \text{kg} \cdot \left(1.34\ \frac{\text{m}}{\text{s}} \right)^2 \right)$$

$E_{Kf} = 1{,}380\ \text{J}$

$E_{Ki} = E_{Kf} = 1{,}380\ \text{J}$

4.

$$m_1 = 375.00\ \text{g} \cdot \frac{1\ \text{kg}}{1000\ \text{g}} = 0.375\ \text{kg}$$

$$m_2 = 5 \cdot 0.375\ \text{kg} = 1.875\ \text{kg}$$

$$v_{1f} = 0.7500\ \frac{\text{m}}{\text{s}}$$

$$v_{2f} = 0.3750\ \frac{\text{m}}{\text{s}}$$

$$v_{1i} = ?$$

$$p_{1i} = (-p_{1f}) + p_{2f}$$

$$p_{1f} = m_1 v_{1f} = 0.375\ \text{kg} \cdot 0.7500\ \frac{\text{m}}{\text{s}} = 0.2813\ \frac{\text{kg} \cdot \text{m}}{\text{s}}$$

$$p_{2f} = m_2 v_{2f} = 1.875\ \text{kg} \cdot 0.3750\ \frac{\text{m}}{\text{s}} = 0.7031\ \frac{\text{kg} \cdot \text{m}}{\text{s}}$$

$$p_{1i} = 0.7031\ \frac{\text{kg} \cdot \text{m}}{\text{s}} - 0.2813\ \frac{\text{kg} \cdot \text{m}}{\text{s}} = 0.4218\ \frac{\text{kg} \cdot \text{m}}{\text{s}}$$

$$p_{1i} = m_1 v_{1i}$$

$$v_{1i} = \frac{p_{1i}}{m_1} = \frac{0.4218\ \frac{\text{kg} \cdot \text{m}}{\text{s}}}{0.375\ \text{kg}} = 1.125\ \frac{\text{m}}{\text{s}}$$

6.

$m_1 = 5.50 \text{ kg}$

$m_2 = 550.0 \text{ g} \cdot \dfrac{1 \text{ kg}}{1000 \text{ g}} = 0.5500 \text{ kg}$

$v_{1f} = 3.2727 \ \dfrac{\text{cm}}{\text{s}} \cdot \dfrac{1 \text{ m}}{100 \text{ cm}} = 0.032727 \ \dfrac{\text{m}}{\text{s}}$

$v_{2f} = 7.2727 \ \dfrac{\text{cm}}{\text{s}} \cdot \dfrac{1 \text{ m}}{100 \text{ cm}} = 0.072727 \ \dfrac{\text{m}}{\text{s}}$

$v_{1i} = ?$

$p_{1i} = p_{1f} + p_{2f}$

$p_{1f} = m_1 v_{1f} = 5.50 \text{ kg} \cdot 0.032727 \ \dfrac{\text{m}}{\text{s}} = 0.18000 \ \dfrac{\text{kg} \cdot \text{m}}{\text{s}}$

$p_{2f} = m_2 v_{2f} = 0.5500 \text{ kg} \cdot 0.072727 \ \dfrac{\text{m}}{\text{s}} = 0.040000 \ \dfrac{\text{kg} \cdot \text{m}}{\text{s}}$

$p_{1i} = 0.18000 \ \dfrac{\text{kg} \cdot \text{m}}{\text{s}} + 0.040000 \ \dfrac{\text{kg} \cdot \text{m}}{\text{s}} = 0.22000 \ \dfrac{\text{kg} \cdot \text{m}}{\text{s}}$

$p_{1i} = m_1 v_{1i}$

$v_{1i} = \dfrac{p_{1i}}{m_1} = \dfrac{0.22000 \ \dfrac{\text{kg} \cdot \text{m}}{\text{s}}}{5.50 \text{ kg}} = 0.0400 \ \dfrac{\text{m}}{\text{s}} \cdot \dfrac{100 \text{ cm}}{1 \text{ m}} = 4.00 \ \dfrac{\text{cm}}{\text{s}}$

7.

$m_1 = 5.50 \text{ kg}$

$m_2 = 0.5500 \text{ kg}$

$v_{1i} = 0.0400 \ \dfrac{\text{m}}{\text{s}}$

$v_{2i} = 0 \text{ J}$

$v_{1f} = 0.032727 \ \dfrac{\text{m}}{\text{s}}$

$v_{2f} = 0.072727 \ \dfrac{\text{m}}{\text{s}}$

$E_{Ki} = ?$

$E_{Kf} = ?$

$$E_{Ki} = \frac{1}{2} m_1 v_{1i}^{\ 2} = \frac{1}{2} \cdot 5.50 \text{ kg} \cdot \left(0.0400 \ \frac{\text{m}}{\text{s}} \right)^2 = 0.00440 \text{ J}$$

$$E_{Kf} = \frac{1}{2} m_1 v_{1f}^{\ 2} + \frac{1}{2} m_2 v_{2f}^{\ 2} = \left(\frac{1}{2} \cdot 5.50 \text{ kg} \cdot \left(0.032727 \ \frac{\text{m}}{\text{s}} \right)^2 \right)$$

$$+ \left(\frac{1}{2} \cdot 0.5500 \text{ kg} \cdot \left(0.072727 \ \frac{\text{m}}{\text{s}} \right)^2 \right) = 0.00440 \text{ J}$$

$E_{Ki} = E_{Kf} = 0.00440 \text{ J}$

8.

$$m_1 = 144.3 \text{ g} \cdot \frac{1 \text{ kg}}{1000 \text{ g}} = 0.1443 \text{ kg}$$

$$m_2 = 10 \cdot 0.1443 \text{ kg} = 1.443 \text{ kg}$$

$$v_{1i} = 65.00 \ \frac{\text{mi}}{\text{hr}} \cdot \frac{5{,}280 \text{ ft}}{1 \text{ mi}} \cdot \frac{0.3048 \text{ m}}{1 \text{ ft}} \frac{1 \text{ hr}}{3{,}600 \text{ s}} = 29.06 \ \frac{\text{m}}{\text{s}}$$

$$v_{1f} = 53.1818 \ \frac{\text{mi}}{\text{hr}} \cdot \frac{5{,}280 \text{ ft}}{1 \text{ mi}} \cdot \frac{0.3048 \text{ m}}{1 \text{ ft}} \frac{1 \text{ hr}}{3{,}600 \text{ s}} = 23.7744 \ \frac{\text{m}}{\text{s}}$$

$$v_{2f} = ?$$

$$p_{1i} = (-p_{1f}) + p_{2f}$$

$$p_{2f} = p_{1i} + p_{1f}$$

$$p_{1f} = m_1 v_{1f} = 0.1443 \text{ kg} \cdot 23.7744 \ \frac{\text{m}}{\text{s}} = 3.4306 \ \frac{\text{kg} \cdot \text{m}}{\text{s}}$$

$$p_{1i} = m_1 v_{1i} = 0.1443 \text{ kg} \cdot 29.06 \ \frac{\text{m}}{\text{s}} = 4.1934 \ \frac{\text{kg} \cdot \text{m}}{\text{s}}$$

$$p_{2f} = 4.1934 \ \frac{\text{kg} \cdot \text{m}}{\text{s}} + 3.4306 \ \frac{\text{kg} \cdot \text{m}}{\text{s}} = 7.624 \ \frac{\text{kg} \cdot \text{m}}{\text{s}}$$

$$p_{2f} = m_2 v_{2f}$$

$$v_{2f} = \frac{p_{2f}}{m_2} = \frac{7.624 \ \frac{\text{kg} \cdot \text{m}}{\text{s}}}{1.443 \text{ kg}} = 5.2834 \ \frac{\text{m}}{\text{s}} \cdot \frac{1 \text{ ft}}{0.3048 \text{ m}} \cdot \frac{1 \text{ mi}}{5{,}280 \text{ ft}} \cdot \frac{3{,}600 \text{ s}}{1 \text{ hr}} = 11.82 \ \frac{\text{mi}}{\text{hr}}$$

Chapter 6

Volume, Mass, and Weight Exercises

1.

$$98.34\ \frac{\text{kg}}{\text{m}^3}\cdot\frac{1000\ \text{g}}{1\ \text{kg}}\cdot\left(\frac{1\ \text{m}}{100\ \text{cm}}\right)^3 = 0.09834\ \frac{\text{g}}{\text{cm}^3}$$

2.

$$42\ \text{mL}\cdot\frac{1\ \text{L}}{1000\ \text{mL}}\cdot\frac{1\ \text{gal}}{3.786\ \text{L}} = 0.011\ \text{gal}$$

3.

$$F_w = 18.5\ \text{lb}\cdot\frac{4.45\ \text{N}}{1\ \text{lb}} = 82.3\ \text{N}$$

$$m = ?$$

$$F_w = mg$$

$$m = \frac{F_w}{g} = \frac{82.3\ \text{N}}{9.80\ \frac{\text{m}}{\text{s}^2}} = 8.40\ \text{kg}$$

4.

$$3.6711\times 10^4\ \frac{\text{g}}{\text{mL}}\cdot\frac{1\ \text{kg}}{1000\ \text{g}}\cdot\frac{1000\ \text{mL}}{1\ \text{L}}\cdot\frac{1000\ \text{L}}{1\ \text{m}^3} = 3.6711\times 10^7\ \frac{\text{kg}}{\text{m}^3}$$

5.

$$1.957\times 10^4\ \text{in}^3\cdot\left(\frac{2.54\ \text{cm}}{1\ \text{in}}\right)^3 = 320{,}700\ \text{cm}^3$$

6.

$$455\ \text{mL}\cdot\frac{1\ \text{L}}{1000\ \text{mL}}\cdot\frac{1\ \text{m}^3}{1000\ \text{L}} = 0.000455\ \text{m}^3$$

7.

$m = 46{,}000 \text{ kg}$

$F_w = ?$

$$F_w = mg = 46{,}000 \text{ kg} \cdot 9.80 \ \frac{\text{m}}{\text{s}^2} = 450{,}000 \text{ N}$$

$$4.508 \times 10^5 \text{ N} \cdot \frac{1 \text{ lb}}{4.45 \text{ N}} = 1.0 \times 10^5 \text{ lb}$$

8.

$$32.11 \text{ L} \cdot \frac{1000 \text{ cm}^3}{1 \text{ L}} \cdot \left(\frac{1 \text{ in}}{2.54 \text{ cm}} \right)^3 = 1{,}959 \text{ in}^3$$

9.

$$F_w = 14.89 \text{ N} \cdot \frac{1 \text{ lb}}{4.45 \text{ N}} = 3.35 \text{ lb}$$

$m = ?$

$$m = \frac{F_w}{g} = \frac{14.89 \text{ N}}{9.80 \ \frac{\text{m}}{\text{s}^2}} = 1.52 \text{ kg}$$

10.

$$36.00 \text{ cm}^3 \cdot \left(\frac{1 \text{ m}}{100 \text{ cm}} \right)^3 = 3.6 \times 10^{-5} \text{ m}^3$$

11.

$$9.11 \text{ m}^3 \cdot \left(\frac{100 \text{ cm}}{1 \text{ m}} \right)^3 = 9.11 \times 10^6 \text{ cm}^3$$

12.

$$4.11 \times 10^5 \text{ m}^3 \cdot \frac{1000 \text{ L}}{1 \text{ m}^3} = 4.11 \times 10^8 \text{ L}$$

13.

$$F_w = 55{,}789 \text{ lb} \cdot \frac{4.45 \text{ N}}{1 \text{ lb}} = 2.48261 \times 10^5 \text{ N}$$

$$m = \frac{F_w}{g} = \frac{2.48261 \times 10^5 \text{ N}}{9.80 \ \frac{\text{m}}{\text{s}^2}} = 25{,}300 \text{ kg}$$

14.

$$5.022\ \frac{\text{g}}{\text{cm}^3}\cdot\frac{1\text{ kg}}{1000\text{ g}}\cdot\left(\frac{100\text{ cm}}{1\text{ m}}\right)^3=5{,}022\ \frac{\text{kg}}{\text{m}^3}$$

15.

$$F_w=50{,}000\text{ N}\cdot\frac{1\text{ lb}}{4.45\text{ N}}=10{,}000\text{ lb}$$

$$m=\frac{F_w}{g}=\frac{50{,}000\text{ N}}{9.80\ \frac{\text{m}}{\text{s}^2}}=5{,}000\text{ kg}$$

16.

$$1.75\times10^{-6}\text{ m}^3\cdot\left(\frac{100\text{ cm}}{1\text{ m}}\right)^3=1.75\text{ cm}^3$$

17.

$$100.5\text{ ft}^3\cdot\left(\frac{12\text{ in}}{1\text{ ft}}\right)^3\cdot\left(\frac{2.54\text{ cm}}{1\text{ in}}\right)^3\cdot\left(\frac{1\text{ m}}{100\text{ cm}}\right)^3=2.846\text{ m}^3$$

18.

$$37\text{ m}^3\cdot\left(\frac{100\text{ cm}}{1\text{ m}}\right)^3\cdot\left(\frac{1\text{ in}}{2.54\text{ cm}}\right)^3=2{,}300{,}000\text{ in}^3$$

19.

$$750\text{ cm}^3\cdot\frac{1\text{ L}}{1000\text{ cm}^3}=0.75\text{ L}$$

20.

$$5{,}755{,}000\text{ gal}\cdot\frac{3.786\text{ L}}{1\text{ gal}}\cdot\frac{1000\text{ cm}^3}{1\text{ L}}\cdot\left(\frac{1\text{ m}}{100\text{ cm}}\right)^3=21{,}790\text{ m}^3$$

Density Exercises

1.

$m = 0.196\ \text{g}$

$V = 100.1\ \text{mL}$

$\rho = ?$

$$\rho = \frac{m}{V} = \frac{0.196\ \text{g}}{100.1\ \text{mL}} = 1.96 \times 10^{-3}\ \frac{\text{g}}{\text{mL}}$$

2.

$$\rho = 955\ \frac{\text{kg}}{\text{m}^3} \cdot \frac{1000\ \text{g}}{1\ \text{kg}} \cdot \left(\frac{1\ \text{m}}{100\ \text{cm}}\right)^3 \cdot \frac{100\ \text{cm}^3}{1\ \text{L}} \cdot \frac{1\ \text{L}}{1000\ \text{mL}} = 0.955\ \frac{\text{g}}{\text{mL}}$$

$m = 550\ \text{g}$

$V = ?$

$$\rho = \frac{m}{V}$$

$$V = \frac{m}{\rho} = \frac{550\ \text{g}}{0.955\ \frac{\text{g}}{\text{mL}}} = 580\ \text{mL}$$

3.

$$m = 15.7\ \text{kg} \cdot \frac{1000\ \text{g}}{1\ \text{kg}} = 1.57 \times 10^4\ \text{g}$$

$$\rho = 5.32\ \frac{\text{g}}{\text{cm}^3}$$

$V = ?$

$$\rho = \frac{m}{V}$$

$$V = \frac{m}{\rho} = \frac{1.57 \times 10^4\ \text{g}}{5.32\ \frac{\text{g}}{\text{cm}^3}} = 2{,}950\ \text{cm}^3$$

$$2.95 \times 10^3\ \text{cm}^3 \cdot \left(\frac{1\ \text{m}}{100\ \text{cm}}\right)^3 = 0.00295\ \text{m}^3$$

4.

$l = 3.00\ \text{cm}$

$w = 3.00\ \text{cm}$

$h = 3.00\ \text{cm}$

$$F_w = 5.336 \times 10^{-2}\ \text{lb} \cdot \frac{4.45\ \text{N}}{1\ \text{lb}} = 2.375 \times 10^{-1}\ \text{N}$$

$\rho = ?$

$F_w = mg$

$$m = \frac{F_w}{g} = \frac{2.375 \times 10^{-1}\ \text{N}}{9.80\ \frac{\text{m}}{\text{s}^2}} = 2.423 \times 10^{-2}\ \text{kg} \cdot \frac{1000\ \text{g}}{1\ \text{kg}} = 24.23\ \text{g}$$

$$V = l \cdot w \cdot h = (3.00\ \text{cm})^3 = 27\ \text{cm}^3$$

$$\rho = \frac{m}{V} = \frac{24.23\ \text{g}}{27\ \text{cm}^3} = 0.897\ \frac{\text{g}}{\text{cm}^3}$$

5.

$V_1 = 23.35\ \text{mL}$

$V_2 = 27.79\ \text{mL}$

$m = 32.1\ \text{g}$

$\rho = ?$

$$V_{Total} = 27.79\ \text{mL} - 23.35\ \text{mL} = 4.44\ \text{mL} \cdot \frac{1\ \text{mL}}{1\ \text{cm}^3} = 4.44\ \text{cm}^3$$

$$\rho = \frac{32.1\ \text{g}}{4.44\ \text{cm}^3} = 7.23\ \frac{\text{g}}{\text{cm}^3}$$

8.

$$\rho = 7{,}830\ \frac{\text{kg}}{\text{m}^3}\cdot\frac{1000\ \text{g}}{1\ \text{kg}}\cdot\left(\frac{1\ \text{m}}{100\ \text{cm}}\right)^3 = 7.83\ \frac{\text{g}}{\text{cm}^3}$$

$l = 2.1\ \text{cm}$

$w = 3.5\ \text{cm}$

$m = 94.5\ \text{g}$

$h = ?$

$$\rho = \frac{m}{V}$$

$$V = \frac{m}{\rho} = \frac{94.5\ \text{g}}{7.83\ \frac{\text{g}}{\text{cm}^3}} = 12.07\ \text{cm}^3$$

$$V = l\cdot w\cdot h$$

$$h = \frac{V}{l\cdot w} = \frac{12.07\ \text{cm}^3}{2.1\ \text{cm}\cdot 3.5\ \text{cm}} = 1.6\ \text{cm}$$

9.

$$m = 306\ \text{g}\cdot\frac{1\ \text{kg}}{1000\ \text{g}} = 0.306\ \text{kg}$$

$$V = 22.5\ \text{mL}\cdot\frac{1\ \text{cm}^3}{1\ \text{mL}}\cdot\left(\frac{1\ \text{m}}{100\ \text{cm}}\right)^3 = 2.25\times10^{-5}\ \text{m}^3$$

$\rho = ?$

$$\rho = \frac{m}{V} = \frac{0.306\ \text{kg}}{2.25\times10^{-5}\ \text{m}^3} = 13{,}600\ \frac{\text{kg}}{\text{m}^3}$$

10.

$$\rho = 1{,}000.0\ \frac{\text{kg}}{\text{m}^3}$$

$$V = 5.6\ \text{L}\cdot\frac{1000\ \text{cm}^3}{1\ \text{L}}\cdot\left(\frac{1\ \text{m}}{100\ \text{cm}}\right)^3 = 5.6\times10^{-3}\ \text{m}^3$$

$m = ?$

$$\rho = \frac{m}{V}$$

$$m = V\rho = 5.6\times10^{-3}\ \text{m}^3\cdot 1{,}000.0\ \frac{\text{kg}}{\text{m}^3} = 5.6\ \text{kg}$$

11.

$$V = 3.0\times10^6 \text{ gal}\cdot\frac{3.786 \text{ L}}{1 \text{ gal}}\cdot\frac{1 \text{ m}^3}{1000 \text{ L}} = 1.14\times10^4 \text{ m}^3$$

$$\rho = 998 \ \frac{\text{kg}}{\text{m}^3}$$

$$F_w = ?$$

$$\rho = \frac{m}{V}$$

$$m = V\rho = 1.14\times10^4 \text{ m}^3\cdot 998 \ \frac{\text{kg}}{\text{m}^3} = 1.13\times10^7 \text{ kg}$$

$$F_w = mg = 1.13\times10^7 \text{ kg}\cdot 9.80 \ \frac{\text{m}}{\text{s}^2} = 1.11\times10^8 \text{ N}$$

$$F_w = 1.11\times10^8 \text{ N}\cdot\frac{1 \text{ lb}}{4.45 \text{ N}} = 25{,}000{,}000 \text{ lb}$$

13.

$$l = 50.0 \text{ m}$$

$$w = 25.0 \text{ m}$$

$$d = 2.00 \text{ m}$$

$$\rho = 998 \ \frac{\text{kg}}{\text{m}^3}$$

$$V = ?$$

$$F_w = ?$$

$$V = lwd = 50.0 \text{ m}\cdot 25.0 \text{ m}\cdot 2.00 \text{ m} = 2{,}500.0 \text{ m}^3\cdot\frac{1000 \text{ L}}{1 \text{ m}^3}\cdot\frac{1 \text{ gal}}{3.786 \text{ L}} = 6.60\times10^5 \text{ gal}$$

$$\rho = \frac{m}{V}$$

$$m = V\rho = 2{,}500.0 \text{ m}^3\cdot 998 \ \frac{\text{kg}}{\text{m}^3} = 2.495\times10^6 \text{ kg}$$

$$F_w = mg = 2.495\times10^6 \text{ kg}\cdot 9.80 \ \frac{\text{m}}{\text{s}^2} = 2.4451\times10^7 \text{ N}\cdot\frac{1 \text{ lb}}{4.45 \text{ N}}\cdot\frac{1 \text{ ton}}{2{,}000 \text{ lb}} = 2{,}750 \text{ tons}$$

Chapter 7

Temperature Unit Conversions

1.a.

$$T_C = 32.0°\text{C}$$

$$T_F = ?$$

$$T_C = \frac{5}{9}(T_F - 32°)$$

$$T_F = \frac{9}{5}T_C + 32° = \frac{9}{5} \cdot 32.0°\text{C} + 32° = 89.6°\text{F}$$

1.b.

$$T_C = 32.0°\text{C}$$

$$T_K = ?$$

$$T_K = T_C + 273.2$$

$$T_K = 32.0°\text{C} + 273.2 = 305.2 \text{ K}$$

2.a.

$$T_F = 56.5°\text{F}$$

$$T_C = ?$$

$$T_C = \frac{5}{9}(T_F - 32°)$$

$$T_C = \frac{5}{9}(56.5°\text{F} - 32°) = 13.6°\text{C}$$

2.b.

$$T_F = 56.5°\text{F}$$

$$T_C = 13.6°\text{C}$$

$$T_K = ?$$

$$T_K = T_C + 273.2$$

$$T_K = 13.6°\text{C} + 273.2 = 286.8 \text{ K}$$

3.a.

$T_K = 455.0\text{ K}$

$T_C = 181.8°\text{C}$

$T_F = ?$

$$T_C = \frac{5}{9}(T_F - 32°)$$

$$T_F = \frac{9}{5}T_C + 32°$$

$$T_F = \frac{9}{5}\cdot 181.8°\text{C} + 32° = 359.2°\text{F}$$

3.b.

$T_K = 455.0\text{ K}$

$T_C = ?$

$T_K = T_C + 273.2$

$T_C = T_K - 273.2 = 455.0\text{ K} - 273.2 = 181.8°\text{C}$

4.a.

$T_F = -17.9°\text{F}$

$T_C = ?$

$$T_C = \frac{5}{9}(T_F - 32°)$$

$$T_C = \frac{5}{9}(-17.9°\text{F} - 32°) = -27.7°\text{C}$$

4.b.

$T_F = -17.9°\text{F}$

$T_C = -27.7°\text{C}$

$T_K = ?$

$T_K = T_C + 273.2$

$T_K = -27.7°\text{C} + 273.2 = 245.5\text{ K}$

5.a.

$$T_C = -41.6°\text{C}$$

$$T_F = ?$$

$$T_C = \frac{5}{9}(T_F - 32°)$$

$$T_F = \frac{9}{5}T_C + 32°$$

$$T_F = \left(\frac{9}{5}\cdot(-41.6°\text{C})\right) + 32° = -42.9°\text{F}$$

5.b.

$$T_C = -41.6°\text{C}$$

$$T_K = ?$$

$$T_K = T_C + 273.2$$

$$T_K = -41.6°\text{C} + 273.2 = 231.6\ \text{K}$$

6.a.

$$T_K = 79.0\ \text{K}$$

$$T_C = ?$$

$$T_K = T_C + 273.2$$

$$T_C = T_K - 273.2$$

$$T_C = 79.0\ \text{K} - 273.2 = -194.2°\text{C}$$

6.b.

$$T_K = 79.0\ \text{K}$$

$$T_C = -194.2°\text{C}$$

$$T_F = ?$$

$$T_C = \frac{5}{9}(T_F - 32°)$$

$$T_F = \frac{9}{5}T_C + 32°$$

$$T_F = \left(\frac{9}{5}\cdot(-194.2°\text{C})\right) + 32° = -317.6°\ \text{F}$$

Chapter 8

Pressure Problems

1.

$$h = 33.5\ \text{ft} \cdot \frac{0.3048\ \text{m}}{1\ \text{ft}} = 10.2\ \text{m}$$

$$\rho = 998\ \frac{\text{kg}}{\text{m}^3}$$

$$P = ?$$

$$P = \rho g h = 998\ \frac{\text{kg}}{\text{m}^3} \cdot 9.80\ \frac{\text{m}}{\text{s}^2} \cdot 10.2\ \text{m} = 99{,}760\ \text{Pa} \cdot \frac{1\ \text{kPa}}{1000\ \text{Pa}} = 99.8\ \text{kPa}$$

$$99{,}760\ \text{Pa} \cdot \frac{14.7\ \text{psi}}{101{,}325\ \text{Pa}} = 14.5\ \text{psi}$$

2.

$$h = 125\ \text{ft} \cdot \frac{0.3048\ \text{m}}{1\ \text{ft}} = 38.1\ \text{m}$$

$$\rho = 1.025\ \frac{\text{g}}{\text{cm}^3} \cdot \frac{1\ \text{kg}}{1000\ \text{g}} \cdot \left(\frac{100\ \text{cm}}{1\ \text{m}}\right)^3 = 1{,}025\ \frac{\text{kg}}{\text{m}^3}$$

$$P = ?$$

$$P = \rho g h = 1{,}025\ \frac{\text{kg}}{\text{m}^3} \cdot 9.80\ \frac{\text{m}}{\text{s}^2} \cdot 38.1\ \text{m} = 383{,}000\ \text{Pa} = 383\ \text{kPa}$$

$$382{,}715\ \text{Pa} \cdot \frac{14.7\ \text{psi}}{101{,}325\ \text{Pa}} = 55.5\ \text{psi}$$

3.

$$\rho = 998\ \frac{\text{kg}}{\text{m}^3}$$

$$P = 101{,}325\ \text{Pa}$$

$$h = ?$$

$$P = \rho g h$$

$$h = \frac{P}{\rho g} = \frac{101{,}325\ \text{Pa}}{998\ \frac{\text{kg}}{\text{m}^3} \cdot 9.80\ \frac{\text{m}}{\text{s}^2}} = 10.4\ \text{m} \cdot \frac{1\ \text{ft}}{0.3048\ \text{m}} = 34.0\ \text{ft}$$

4.

$$h = 60.6\ \text{ft} \cdot \frac{0.3048\ \text{m}}{1\ \text{ft}} = 18.47\ \text{m}$$

$$\rho = 998\ \frac{\text{kg}}{\text{m}^3}$$

$$P = ?$$

$$P = \rho g h = 998\ \frac{\text{kg}}{\text{m}^3} \cdot 9.80\ \frac{\text{m}}{\text{s}^2} \cdot 18.47\ \text{m} = 181\ \text{kPa}$$

5.

$$h = 18.47\ \text{m}$$

$$D = 10.0\ \text{ft} \cdot \frac{0.3048\ \text{m}}{1\ \text{ft}} = 3.048\ \text{m}$$

$$\rho = 998\ \frac{\text{kg}}{\text{m}^3}$$

$$P = ?$$

$$V = \pi r^2 h = \pi \cdot \left(\frac{3.048\ \text{m}}{2} \right)^2 \cdot 18.47\ \text{m} = 134.8\ \text{m}^3$$

$$\rho = \frac{m}{V}$$

$$m = \rho V = 998\ \frac{\text{kg}}{\text{m}^3} \cdot 134.8\ \text{m}^3 = 134{,}530.4\ \text{kg}$$

$$F_w = mg = 134{,}530.4\ \text{kg} \cdot 9.80\ \frac{\text{m}}{\text{s}^2} = 1{,}318{,}398\ \text{N}$$

$$A = \pi r^2 = \pi \left(\frac{3.048\ \text{m}}{2} \right)^2 = 7.297\ \text{m}^2$$

$$P = \frac{F_w}{A} = \frac{1{,}318{,}398\ \text{N}}{7.297\ \text{m}^2} = 181\ \text{kPa}$$

6.

$$P = 181{,}000\ \text{Pa}$$

$$P_{atm(atmospheric)} = 101{,}325\ \text{Pa}$$

$$P_{abs(absolute)} = ?$$

$$P_{abs} = P_{gauge} + P_{atm} = 181{,}000\ \text{Pa} + 101{,}325\ \text{Pa} = 282\ \text{kPa}$$

$$282{,}000\ \text{Pa} \cdot \frac{14.7\ \text{psi}}{101{,}325\ \text{Pa}} = 40.9\ \text{psi}$$

7.

$P_{gauge} = 55.0\ \text{psi}$

$P_{abs_i} = 40.9\ \text{psi}$

$P_{abs_f} = ?$

$P_{abs_f} = P_{abs_i} + P_{gauge} = 40.9\ \text{psi} + 55.0\ \text{psi} = 95.9\ \text{psi}$

$$95.9\ \text{psi} \cdot \frac{101{,}325\ \text{Pa}}{14.7\ \text{psi}} = 661\ \text{kPa}$$

8.a.

$P_{gauge} = 14.5\ \text{psi}$

$P_{atm} = 14.7\ \text{psi}$

$P_{abs} = P_{gauge} + P_{atm} = 14.5\ \text{psi} + 14.7\ \text{psi} = 29.2\ \text{psi}$

$$29.2\ \text{psi} \cdot \frac{101{,}325\ \text{Pa}}{14.7\ \text{psi}} = 201.2\ \text{kPa}$$

8.b.

$P_{gauge} = 14.5\ \text{psi}$

$P_{atm} = 14.7\ \text{psi}$

$P_{abs} = P_{gauge} + P_{atm} = 55.5\ \text{psi} + 14.7\ \text{psi} = 70.2\ \text{psi}$

$$70.2\ \text{psi} \cdot \frac{101{,}325\ \text{Pa}}{14.7\ \text{psi}} = 484\ \text{kPa}$$

9.

$$P = 83.4\ \text{kPa} \cdot \frac{1000\ \text{Pa}}{1\ \text{kPa}} = 83{,}400\ \text{Pa}$$

$$\rho = 13{,}600\ \frac{\text{kg}}{\text{m}^3}$$

$h = ?$

$P = \rho g h$

$$h = \frac{P}{\rho g} = \frac{83{,}400\ \text{Pa}}{13{,}600\ \frac{\text{kg}}{\text{m}^3} \cdot 9.80\ \frac{\text{m}}{\text{s}^2}} = 0.626\ \text{m} = 626\ \text{mm}$$

10.

$$P = 2{,}800\ \text{psi} \cdot \frac{101{,}325\ \text{Pa}}{14.7\ \text{psi}} = 1.93 \times 10^7\ \text{Pa}$$

$$F_w = 2{,}200\ \text{lb} \cdot \frac{4.45\ \text{N}}{1\ \text{lb}} = 9{,}790\ \text{N}$$

$$D = 0.75\ \text{in} \cdot \frac{2.54\ \text{cm}}{1\ \text{in}} \cdot \frac{1\ \text{m}}{100\ \text{cm}} = 0.01905\ \text{m}$$

$$P_{foot} = ?$$

P_{foot} is the pressure under each foot.

$$r = \frac{D}{2} = \frac{0.01905\ \text{m}}{2} = 0.009525\ \text{m}$$

$$A = \pi r^2 = \pi(0.009525\ \text{m})^2 = 2.850 \times 10^{-4}\ \text{m}^2$$

Sin ce there are four feet, the total pressure will be divided by four.

$$P_{foot} = \frac{F}{4A} = \frac{9{,}790\ \text{N}}{4 \cdot 2.850 \times 10^{-4}\ \text{m}^2} = 8{,}588{,}000\ \text{Pa} \cdot \frac{14.7\ \text{psi}}{101{,}325\ \text{Pa}} = 1{,}200\ \text{psi}$$

11.

$$P = 3{,}200\ \text{psi} \cdot 0.75 = 2{,}400\ \text{psi} \cdot \frac{101{,}325\ \text{Pa}}{14.7\ \text{psi}} = 1.65 \times 10^7\ \text{Pa}$$

$$\rho = 998\ \frac{\text{kg}}{\text{m}^3}$$

$$h = ?$$

$$P = \rho g h$$

$$h = \frac{P}{\rho g} = \frac{1.65 \times 10^7\ \text{Pa}}{998\ \frac{\text{kg}}{\text{m}^3} \cdot 9.80\ \frac{\text{m}}{\text{s}^2}} = 1{,}700\ \text{m}$$

12.

$$A = 120\ \text{cm}^2 \cdot \left(\frac{1\ \text{m}}{100\ \text{cm}}\right)^2 = 0.012\ \text{m}^2$$

$$F_w = 195\ \text{lb} \cdot \frac{4.45\ \text{N}}{1\ \text{lb}} = 867.8\ \text{N}$$

$$P_{shoes} = ?$$

$$P_{shoes} = \frac{F}{2A} = \frac{867.8\ \text{N}}{2 \cdot 0.012\ \text{m}^2} = 36{,}000\ \text{Pa}$$

$$36{,}000\ \text{Pa} \cdot \frac{14.7\ \text{psi}}{101{,}325\ \text{Pa}} = 5.2\ \text{psi}$$

13.

$$l = 0.750\ \text{cm} \cdot \frac{1\ \text{m}}{100\ \text{cm}} = 0.00750\ \text{m}$$

$$w = 0.750\ \text{cm} \cdot \frac{1\ \text{m}}{100\ \text{cm}} = 0.00750\ \text{m}$$

$$F_w = 130\ \text{lb} \cdot \frac{4.45\ \text{N}}{1\ \text{lb}} = 579\ \text{N}$$

$$P_{heels} = ?$$

$$P_{heels} = \frac{F_w}{2A}$$

$$A = lw = (0.00750\ \text{m})^2 = 5.6 \times 10^{-5}\ \text{m}^2$$

$$P_{heels} = \frac{579\ \text{N}}{2 \cdot 5.6 \times 10^{-5}\ \text{m}^2} = 5{,}200{,}000\ \text{Pa}$$

$$5{,}200{,}000\ \text{Pa} \cdot \frac{14.7\ \text{psi}}{101{,}325\ \text{Pa}} = 750\ \text{psi}$$

14.

$$h = 20.0\ \text{ft} \cdot \frac{0.3048\ \text{m}}{1\ \text{ft}} = 6.096\ \text{m}$$

$$\rho = 998\ \frac{\text{kg}}{\text{m}^3}$$

$$P_{abs} = ?$$

$$P = \rho g h = 998\ \frac{\text{kg}}{\text{m}^3} \cdot 9.80\ \frac{\text{m}}{\text{s}^2} \cdot 6.096\ \text{m} = 5.96 \times 10^4\ \text{Pa}$$

$$P_{abs} = P_{gauge} + P_{atm} = 5.96 \times 10^4\ \text{Pa} + 101{,}325\ \text{Pa} = 1.609 \times 10^5\ \text{Pa}$$

$$P_{abs} = 1.609 \times 10^5\ \text{Pa} \cdot \frac{14.7\ \text{psi}}{101{,}325\ \text{Pa}} = 23.3\ \text{psi}$$

Buoyancy Problems

1.

$$w = 1.5\ \text{in} \cdot \frac{2.54\ \text{cm}}{1\ \text{in}} \cdot \frac{1\ \text{m}}{100\ \text{cm}} = 0.0381\ \text{m}$$

$$h = 3.5\ \text{in} \cdot \frac{2.54\ \text{cm}}{1\ \text{in}} \cdot \frac{1\ \text{m}}{100\ \text{cm}} = 0.0889\ \text{m}$$

$$l = 12\ \text{in} \cdot \frac{2.54\ \text{cm}}{1\ \text{in}} \cdot \frac{1\ \text{m}}{100\ \text{cm}} = 0.3048\ \text{m}$$

$$\rho = 998\ \frac{\text{kg}}{\text{m}^3}$$

$$F_{B(Buoyant)} = ?$$

$$V = lwh = 0.0381\ \text{m} \cdot 0.0889\ \text{m} \cdot 0.3048\ \text{m} = 1.03 \times 10^{-3}\ \text{m}^3$$

$$\rho = \frac{m}{V}$$

$$m = V\rho = 1.03 \times 10^{-3}\ \text{m}^3 \cdot 998\ \frac{\text{kg}}{\text{m}^3} = 1.03\ \text{kg}$$

$$F_w = mg = 1.03\ \text{kg} \cdot 9.80\ \frac{\text{m}}{\text{s}^2} = 1.0 \times 10^1\ \text{N}$$

$$10.1\ \text{N} \cdot \frac{1\ \text{lb}}{4.45\ \text{N}} = 2.3\ \text{lb}$$

2.

$$V = 1.03\times10^{-3}\ \text{m}^3$$

$$\rho_{saltwater} = 1.025\ \frac{\text{g}}{\text{cm}^3}\cdot\frac{1\ \text{kg}}{1000\ \text{g}}\cdot\left(\frac{100\ \text{cm}}{1\ \text{m}}\right)^3 = 1{,}025\ \frac{\text{kg}}{\text{m}^3}$$

$$\rho_{mercury} = 13.6\ \frac{\text{g}}{\text{cm}^3}\cdot\frac{1\ \text{kg}}{1000\ \text{g}}\cdot\left(\frac{100\ \text{cm}}{1\ \text{m}}\right)^3 = 1.36\times10^4\ \frac{\text{kg}}{\text{m}^3}$$

$$F_{B,\ saltwater} = ?$$

$$F_{B,\ mercury} = ?$$

$$\rho_{saltwater} = \frac{m}{V}$$

$$m = \rho_{saltwater}V = 1{,}025\ \frac{\text{kg}}{\text{m}^3}\cdot 1.03\times10^{-3}\ \text{m}^3 = 1.056\ \text{kg}$$

$$F_{B,\ saltwater} = F_{w,\ saltwater} = mg = 1.056\ \text{kg}\cdot 9.80\ \frac{\text{m}}{\text{s}^2} = 1.0\times10^1\ \text{N}$$

$$F_{B,\ saltwater} = 10.3\ \text{N}\cdot\frac{1\ \text{lb}}{4.45\ \text{N}} = 2.3\ \text{lb}$$

$$\rho_{mercury} = \frac{m}{V}$$

$$m = \rho_{mercury}V = 1.36\times10^4\ \frac{\text{kg}}{\text{m}^3}\cdot 1.03\times10^{-3}\ \text{m}^3 = 14.0\ \text{kg}$$

$$F_{B,\ mercury} = F_{w,\ mercury} = mg = 14.0\ \text{kg}\cdot 9.80\ \frac{\text{m}}{\text{s}^2} = 140\ \text{N}$$

$$F_{B,\ mercury} = 140\ \text{N}\cdot\frac{1\ \text{lb}}{4.45\ \text{N}} = 31\ \text{lb}$$

3.

$$D = 22.5 \text{ in} \cdot \frac{2.54 \text{ cm}}{1 \text{ in}} \cdot \frac{1 \text{ m}}{100 \text{ cm}} = 0.5715 \text{ m}$$

$$h = 33.5 \text{ in} \cdot \frac{2.54 \text{ cm}}{1 \text{ in}} \cdot \frac{1 \text{ m}}{100 \text{ cm}} = 0.8509 \text{ m}$$

$$F_w = 54.0 \text{ lb} \cdot \frac{4.45 \text{ N}}{1 \text{ lb}} = 240.3 \text{ N}$$

$$F_{cable} = ?$$

$$V = \pi \left(\frac{D}{2} \right)^2 h = \pi \left(\frac{0.5715 \text{ m}}{2} \right)^2 \cdot 0.8509 \text{ m} = 0.2187 \text{ m}^3$$

$$\rho = \frac{m}{V}$$

$$m = \rho V = 998 \ \frac{\text{kg}}{\text{m}^3} \cdot 0.2187 \text{ m}^3 = 218.3 \text{ kg}$$

$$F_w = mg = 218.3 \text{ kg} \cdot 9.80 \ \frac{\text{m}}{\text{s}^2} = 2.139 \times 10^3 \text{ N}$$

$$F_B = F_w + F_a$$

$$F_a = F_{cable} = F_B - F_w = 2.139 \times 10^3 \text{ N} - 240.3 \text{ N} = 1.90 \times 10^3 \text{ N}$$

4.

$$\rho_{s(styrofoam)} = 55\ \frac{kg}{m^3}$$

$$h = 12\ in \cdot \frac{2.54\ cm}{1\ in} \cdot \frac{1\ m}{100\ cm} = 0.305\ m$$

$$l = 8.0\ ft \cdot \frac{0.3048\ m}{1\ ft} = 2.44\ m$$

$$w = 6.0\ ft \cdot \frac{0.3048\ m}{1\ ft} = 1.83\ m$$

$$F_{a(additional)} = ?$$

$$V_s = 0.305\ m \cdot 2.44\ m \cdot 1.83\ m = 1.36\ m^3$$

$$\rho_s = \frac{m_s}{V_s}$$

$$m_s = \rho_s V_s = 55\ \frac{kg}{m^3} \cdot\ 1.36\ m^3 = 74.8\ kg$$

$$F_w = m_s g = 74.8\ kg \cdot 9.80\ \frac{m}{s^2} = 733\ N$$

$$V_{dw(displaced\ water)} = \frac{V_{block}}{2} = 0.681\ m^3$$

$$m_{dw} = \rho_{water} V_{dw} = 998\ \frac{kg \cdot m}{s} \cdot 0.681\ m^3 = 679.6\ kg$$

$$F_B = m_{dw} g = 679.6\ kg \cdot 9.80\ \frac{m}{s^2} = 6.66 \times 10^3\ N$$

$$F_B = F_w + F_a$$

$$F_a = F_B - F_w = 6.66 \times 10^3\ N - 733\ N = 5{,}900\ N$$

$$5{,}927\ N \cdot \frac{1\ lb}{4.45\ N} = 1{,}300\ lb$$

5.

$$F_{w,\,dock} = 450\text{ lb}\cdot\frac{4.45\text{ N}}{1\text{ lb}} = 2{,}002\text{ N}$$

$$V_{drums} = 8\cdot 0.219\text{ m}^3 = 1.75\text{ m}^3$$

$$F_{w,\,drums} = 8\cdot 54.0\text{ lb}\cdot\frac{4.45\text{ N}}{1\text{ lb}} = 1{,}922\text{ N}$$

$$F_a = ?$$

$$F_{w,\,total} = F_{w,\,drums} + F_{w,\,dock} = 2{,}002\text{ N} + 1{,}922\text{ N} = 3{,}924\text{ N}$$

$$V_{dw(displaced\ water)} = \frac{V_{drums}}{2} = \frac{1.75\text{ m}^3}{2} = 0.875\text{ m}^3$$

$$\rho_{water} = \frac{m_{dw}}{V_{dw}}$$

$$m_{dw} = V_{dw}\rho_{water} = 0.875\text{ m}^3\cdot 998\ \frac{\text{kg}}{\text{m}^3} = 873\text{ kg}$$

$$F_B = m_{dw}g = 873\text{ kg}\cdot 9.80\ \frac{\text{m}}{\text{s}^2} = 8{,}555\text{ N}$$

$$F_B = F_w + F_a$$

$$F_a = F_B - F_w = 8{,}555\text{ N} - 3{,}924\text{ N} = 4{,}600\text{ N}$$

$$4{,}631\text{ N}\cdot\frac{1\text{ lb}}{4.45\text{ N}} = 1.0\times 10^3\text{ lb}$$

Chapter 9

1.

$$f = 60.0 \text{ Hz}$$

$$\tau = ?$$

$$\tau = \frac{1}{f}$$

$$\tau = \frac{1}{60.0 \text{ Hz}} = 0.0167 \text{ s}$$

2.

$$\tau = 2.155 \times 10^{-5} \text{ s}$$

$$f = ?$$

$$\tau = \frac{1}{f}$$

$$f = \frac{1}{\tau} = \frac{1}{2.155 \times 10^{-5} \text{ s}} = 46.40 \text{ kHz}$$

3.

$$f = 26.0 \text{ kHz} \cdot \frac{1000 \text{ Hz}}{1 \text{ kHz}} = 2.60 \times 10^{4} \text{ Hz}$$

$$\tau = ?$$

$$\tau = \frac{1}{f} = \frac{1}{2.60 \times 10^{4} \text{ Hz}} = 3.85 \times 10^{-5} \text{ s}$$

4.

$$f = 2.60 \times 10^{4} \text{ Hz}$$

$$v = 342 \ \frac{\text{m}}{\text{s}}$$

$$\lambda = ?$$

$$v = \lambda f$$

$$\lambda = \frac{v}{f} = \frac{342 \ \frac{\text{m}}{\text{s}}}{2.60 \times 10^{4} \text{ Hz}} = 1.32 \times 10^{-2} \text{ m} \cdot \frac{100 \text{ cm}}{1 \text{ m}} = 1.32 \text{ cm}$$

5.

$$f = 89.5\ \text{MHz} \cdot \frac{10^6\ \text{Hz}}{1\ \text{MHz}} = 8.95 \times 10^7\ \text{Hz}$$

$$v = c = 3.00 \times 10^8\ \frac{\text{m}}{\text{s}}$$

$$\lambda = ?$$

$$\tau = ?$$

$$v = \lambda f$$

$$\lambda = \frac{v}{f} = \frac{3.00 \times 10^8\ \frac{\text{m}}{\text{s}}}{8.95 \times 10^7\ \text{Hz}} = 3.35\ \text{m}$$

$$\tau = \frac{1}{f} = \frac{1}{8.95 \times 10^7\ \text{Hz}} = 1.12 \times 10^{-8}\ \text{s} \cdot \frac{10^6\ \mu\text{s}}{1\ \text{s}} = 0.0112\ \mu\text{s}$$

6.

$$v = c = 3.00 \times 10^8\ \frac{\text{m}}{\text{s}}$$

$$f = 1{,}310\ \text{kHz} \cdot \frac{1000\ \text{Hz}}{1\ \text{kHz}} = 1.31 \times 10^6\ \text{Hz}$$

$$\lambda = ?$$

$$\tau = ?$$

$$v = \lambda f$$

$$\lambda = \frac{v}{f} = \frac{3.00 \times 10^8\ \frac{\text{m}}{\text{s}}}{1.31 \times 10^6\ \text{Hz}} = 229\ \text{m}$$

$$\tau = \frac{1}{f} = \frac{1}{1.31 \times 10^6\ \text{Hz}} = 7.64 \times 10^{-7}\ \text{s} \cdot \frac{10^6\ \mu\text{s}}{1\ \text{s}} = 0.763\ \mu\text{s}$$

8.

$$v = c = 3.00\times10^{8}\ \frac{\text{m}}{\text{s}}$$

$$\lambda = 542\ \text{nm}\cdot\frac{1\ \text{m}}{10^{9}\ \text{nm}} = 5.42\times10^{-7}\ \text{m}$$

$$\tau = ?$$

$$f = ?$$

$$v = \lambda f$$

$$f = \frac{v}{\lambda} = \frac{3.00\times10^{8}\ \frac{\text{m}}{\text{s}}}{5.42\times10^{-7}\ \text{m}} = 5.54\times10^{14}\ \text{Hz}\cdot\frac{1\ \text{GHz}}{10^{9}\ \text{Hz}} = 5.54\times10^{5}\ \text{GHz}$$

$$\tau = \frac{1}{f} = \frac{1}{5.54\times10^{14}\ \text{Hz}} = 1.81\times10^{-15}\ \text{s}\cdot\frac{10^{9}\ \text{ns}}{1\ \text{s}} = 1.81\times10^{-6}\ \text{ns}$$

9.

$$\lambda = 10.6\ \mu\text{m}\cdot\frac{1\ \text{m}}{10^{6}\ \mu\text{m}} = 1.06\times10^{-5}\ \text{m}$$

$$v = 3.00\times10^{8}\ \frac{\text{m}}{\text{s}}$$

$$\tau = ?$$

$$f = ?$$

$$v = \lambda f$$

$$f = \frac{v}{\lambda} = \frac{3.00\times10^{8}\ \frac{\text{m}}{\text{s}}}{1.06\times10^{-5}\ \text{m}} = 2.83\times10^{13}\ \text{Hz}\cdot\frac{1\ \text{MHz}}{10^{6}\ \text{Hz}} = 2.83\times10^{7}\ \text{MHz}$$

$$\tau = \frac{1}{f} = \frac{1}{2.83\times10^{13}\ \text{Hz}} = 3.53\times10^{-14}\ \text{s}\cdot\frac{10^{6}\mu\text{s}}{1\ \text{s}} = 3.53\times10^{-8}\ \mu\text{s}$$

10.

$$f = 33\ \text{kHz} \cdot \frac{1000\ \text{Hz}}{1\ \text{kHz}} = 3.3 \times 10^4\ \text{Hz}$$

$$v = 342\ \frac{\text{m}}{\text{s}}$$

$$\tau = ?$$

$$\lambda = ?$$

$$\tau = \frac{1}{f} = \frac{1}{3.3 \times 10^4\ \text{Hz}} = 3.03 \times 10^{-5}\ \text{s} \cdot \frac{10^3\ \text{ms}}{1\ \text{s}} = 0.030\ \text{ms}$$

$$v = \lambda f$$

$$\lambda = \frac{v}{f} = \frac{342\ \frac{\text{m}}{\text{s}}}{3.3 \times 10^4\ \text{Hz}} = 1.04 \times 10^{-2}\ \text{m} \cdot \frac{1000\ \text{mm}}{1\ \text{m}} = 1.0 \times 10^1\ \text{mm}$$

11.

$$v = 3.00 \times 10^8\ \frac{\text{m}}{\text{s}}$$

$$\lambda = 2.00\ \text{mm} \cdot \frac{1\ \text{m}}{10^3\ \text{mm}} = 2.00 \times 10^{-3}\ \text{m}$$

$$v = \lambda f$$

$$f = \frac{v}{\lambda} = \frac{3.00 \times 10^8\ \frac{\text{m}}{\text{s}}}{2.00 \times 10^{-3}\ \text{m}} = 1.50 \times 10^{11}\ \text{Hz}$$

12.

$$\lambda_b = 470 \text{ nm} \cdot \frac{1 \text{ m}}{10^9 \text{ nm}} = 4.7 \times 10^{-7} \text{ m}$$

$$\lambda_g = 550 \text{ nm} \cdot \frac{1 \text{ m}}{10^9 \text{ nm}} = 5.5 \times 10^{-7} \text{ m}$$

$$\lambda_r = 680 \text{ nm} \cdot \frac{1 \text{ m}}{10^9 \text{ nm}} = 6.8 \times 10^{-7} \text{ m}$$

$$v = 3.00 \times 10^8 \ \frac{\text{m}}{\text{s}}$$

$$f_b = ?$$

$$f_g = ?$$

$$f_r = ?$$

$$v = \lambda f$$

$$f = \frac{v}{\lambda}$$

$$f_b = \frac{3.00 \times 10^8 \ \frac{\text{m}}{\text{s}}}{4.7 \times 10^{-7} \text{ m}} = 6.4 \times 10^{14} \text{ Hz}$$

$$f_g = \frac{3.00 \times 10^8 \ \frac{\text{m}}{\text{s}}}{5.5 \times 10^{-7} \text{ m}} = 5.5 \times 10^{14} \text{ Hz}$$

$$f_r = \frac{3.00 \times 10^8 \ \frac{\text{m}}{\text{s}}}{6.8 \times 10^{-7} \text{ m}} = 4.4 \times 10^{14} \text{ Hz}$$

13.

$$f = 20.00 \text{ Hz}$$

$$v = 342.0 \ \frac{\text{m}}{\text{s}}$$

$$\lambda = ?$$

$$v = \lambda f$$

$$\lambda = \frac{v}{f} = \frac{342.0 \ \frac{\text{m}}{\text{s}}}{20.00 \text{ Hz}} = 17.10 \text{ m}$$

14.a.

$$f = 1.00 \text{ kHz} \cdot \frac{1000 \text{ Hz}}{1 \text{ kHz}} = 1.00 \times 10^3 \text{ Hzs}$$

$$v = 342 \ \frac{\text{m}}{\text{s}}$$

$$\lambda = ?$$

$$v = \lambda f$$

$$\lambda = \frac{v}{f} = \frac{342 \ \frac{\text{m}}{\text{s}}}{1.00 \times 10^3 \text{ Hz}} = 0.342 \text{ m}$$

14.b.

$$f = 1.00 \text{ kHz} \cdot \frac{1000 \text{ Hz}}{1 \text{ kHz}} = 1.00 \times 10^3 \text{ Hz}$$

$$v = 1{,}402 \ \frac{\text{m}}{\text{s}}$$

$$\lambda = ?$$

$$v = \lambda f$$

$$\lambda = \frac{v}{f} = \frac{1{,}402 \ \frac{\text{m}}{\text{s}}}{1.00 \times 10^3 \text{ Hz}} = 1.40 \text{ m}$$

14.c.

$$f = 1.00 \text{ kHz} \cdot \frac{1000 \text{ Hz}}{1 \text{ kHz}} = 1.00 \times 10^3 \text{ Hz}$$

$$v = 5{,}130 \ \frac{\text{m}}{\text{s}}$$

$$\lambda = ?$$

$$v = \lambda f$$

$$\lambda = \frac{v}{f} = \frac{5{,}130 \ \frac{\text{m}}{\text{s}}}{1.00 \times 10^3 \text{ Hz}} = 5.13 \text{ m}$$

14.d.

$$f = 1.00 \text{ kHz} \cdot \frac{1000 \text{ Hz}}{1 \text{ kHz}} = 1.00 \times 10^3 \text{ Hz}$$

$$v = 965 \ \frac{\text{m}}{\text{s}}$$

$$\lambda = ?$$

$$v = \lambda f$$

$$\lambda = \frac{v}{f} = \frac{965 \ \frac{\text{m}}{\text{s}}}{1.00 \times 10^3 \text{ Hz}} = 0.965 \text{ m}$$

15.a.

$$f = 4.67 \times 10^{20} \text{ Hz}$$

$$v = 3.00 \times 10^8 \ \frac{\text{m}}{\text{s}}$$

$$\lambda = ?$$

$$v = \lambda f$$

$$\lambda = \frac{v}{f} = \frac{3.00 \times 10^8 \ \frac{\text{m}}{\text{s}}}{4.67 \times 10^{20} \text{ Hz}} = 6.42 \times 10^{-13} \text{ m}$$

$$6.42 \times 10^{-13} \text{ m} \cdot \frac{10^9 \text{ nm}}{1 \text{ m}} = 0.000642 \text{ nm}$$

15.b.

$$f = 9.9876 \times 10^{18} \text{ Hz}$$

$$v = 3.00 \times 10^8 \ \frac{\text{m}}{\text{s}}$$

$$\lambda = ?$$

$$v = \lambda f$$

$$\lambda = \frac{v}{f} = \frac{3.00 \times 10^8 \ \frac{\text{m}}{\text{s}}}{9.9876 \times 10^{18} \text{ Hz}} = 3.00 \times 10^{-11} \text{ m}$$

$$3.00 \times 10^{-11} \text{ m} \cdot \frac{10^9 \text{ nm}}{1 \text{ m}} = 0.0300 \text{ nm}$$

15.c.

$f = 2.555 \times 10^{10}$ Hz

$v = 3.00 \times 10^{8}\ \dfrac{\text{m}}{\text{s}}$

$\lambda = ?$

$v = \lambda f$

$$\lambda = \frac{v}{f} = \frac{3.00 \times 10^{8}\ \dfrac{\text{m}}{\text{s}}}{2.555 \times 10^{10}\ \text{Hz}} = 1.17 \times 10^{-2}\ \text{m}$$

$$1.17 \times 10^{-2}\ \text{m} \cdot \frac{10^{9}\ \text{nm}}{1\ \text{m}} = 11{,}700{,}000\ \text{nm}$$

15.d.

$f = 1.172 \times 10^{15}$ Hz

$v = 3.00 \times 10^{8}\ \dfrac{\text{m}}{\text{s}}$

$\lambda = ?$

$v = \lambda f$

$$\lambda = \frac{v}{f} = \frac{3.00 \times 10^{8}\ \dfrac{\text{m}}{\text{s}}}{1.172 \times 10^{15}\ \text{Hz}} = 2.56 \times 10^{-7}\ \text{m}$$

$$2.56 \times 10^{-7}\ \text{m} \cdot \frac{10^{9}\ \text{nm}}{1\ \text{m}} = 256\ \text{nm}$$

15.e.

$f = 2.83 \times 10^{13}$ Hz

$v = 3.00 \times 10^{8}\ \dfrac{\text{m}}{\text{s}}$

$\lambda = ?$

$v = \lambda f$

$$\lambda = \frac{v}{f} = \frac{3.00 \times 10^{8}\ \dfrac{\text{m}}{\text{s}}}{2.83 \times 10^{13}\ \text{Hz}} = 1.06 \times 10^{-5}\ \text{m}$$

$$1.06 \times 10^{-5}\ \text{m} \cdot \frac{10^{9}\ \text{nm}}{1\ \text{m}} = 10{,}600\ \text{nm}$$

16.

$$f = 2.45 \times 10^9 \text{ Hz}$$

$$D = 3 \text{ mm} \cdot \frac{1 \text{ m}}{10^3 \text{ mm}} = 0.003 \text{ m}$$

$$ratio = ?$$

$$\lambda = ?$$

$$v = \lambda f$$

$$\lambda = \frac{v}{f} = \frac{3.00 \times 10^8 \ \frac{\text{m}}{\text{s}}}{2.45 \times 10^9 \text{ Hz}} = 0.1224 \text{ m}$$

$$\frac{\lambda}{D} = \frac{0.12 \text{ m}}{0.003 \text{ m}} = 40$$

Chapter 11

Introductory Circuit Calculations

1.

$I = 13.00 \text{ A}$

$V = 25.00 \text{ V}$

$R = ?$

$V = IR$

$$R = \frac{V}{I} = \frac{25.00 \text{ V}}{13.00 \text{ A}} = 1.923 \ \Omega$$

2.

$V = 24 \text{ V}$

$R = 250 \ \Omega$

$I = ?$

$V = IR$

$$I = \frac{V}{R} = \frac{24 \text{ V}}{250 \ \Omega} = 0.096 \text{ A} \cdot \frac{1000 \text{ mA}}{1 \text{ A}} = 96 \text{ mA}$$

3.

$R = 12.20 \text{ k}\Omega$

$V = 4.500 \text{ V}$

$I = ?$

$V = IR$

$$I = \frac{V}{R} = \frac{4.500 \text{ V}}{12.20 \text{ k}\Omega} = 0.3689 \text{ mA}$$

4.

$I = 0.0300 \text{ mA}$

$$R = 33.3 \text{ M}\Omega \cdot \frac{10^3 \text{ k}\Omega}{1 \text{ M}\Omega} = 3.33 \times 10^4 \text{ k}\Omega$$

$V = ?$

$V = IR = 0.0300 \text{ mA} \cdot 3.33 \times 10^4 \text{ k}\Omega = 999 \text{ V}$

5.a.

$I = 13.00\ \text{A}$

$V = 25.00\ \text{V}$

$P = ?$

$P = VI = 25.00\ \text{V} \cdot 13.00\ \text{A} = 325.0\ \text{W}$

5.b.

$V = 24\ \text{V}$

$I = 96\ \text{mA}$

$P = ?$

$P = VI = 24\ \text{V} \cdot 96\ \text{mA} = 2{,}300\ \text{mW} = 2.3\ \text{W}$

5.c.

$V = 4.500\ \text{V}$

$I = 0.3689\ \text{mA}$

$P = ?$

$P = VI = 4.500\ \text{V} \cdot 0.3689\ \text{mA} = 1.660\ \text{mW}$

6.

$V = 120\ \text{V}$

$P = 60.00\ \text{W}$

$R = ?$

$I = ?$

$P = VI$

$$I = \frac{P}{V} = \frac{60.00\ \text{W}}{120\ \text{V}} = 0.50\ \text{A}$$

$V = IR$

$$R = \frac{V}{I} = \frac{120\ \text{V}}{0.50\ \text{A}} = 240\ \Omega$$

7.

$V = 120\ \text{V}$

$I = 12\ \text{A}$

$P = ?$

$$P = VI = 120\ \text{V} \cdot 12\ \text{A} = 1{,}440\ \text{W} \cdot \frac{1\ \text{kW}}{1000\ \text{W}} = 1.4\ \text{kW}$$

8.

$I = 13.5\ \mu\text{A}$

$V = 6.0\ \text{V}$

$P = ?$

$P = VI = 6.0\ \text{V} \cdot 13.5\ \mu\text{A} = 81\ \mu\text{W}$

9.

$$P = 155\ \text{MW} \cdot \frac{10^6\ \text{W}}{1\ \text{MW}} = 1.55 \times 10^8\ \text{W}$$

$V = 762\ \text{V}$

$I = ?$

$P = VI$

$$I = \frac{P}{V} = \frac{1.55 \times 10^8\ \text{W}}{762\ \text{V}} = 203{,}000\ \text{A}$$

Equivalent Resistance Calculations

1.

$R_3 + R_4 = 47\ \Omega\ + 130\ \Omega\ = 177\ \Omega$

$$R_2 \,||\, 177\ \Omega\ = \frac{32\ \Omega\ \cdot 177\ \Omega}{32\ \Omega\ + 177\ \Omega} = 27.1005\ \Omega$$

$R_{EQ} = R_1 + 27.1005\ \Omega\ _4 = 55\ \Omega\ + 27.1005\ \Omega\ = 82.1005\ \Omega$

2.

$$R_3 \,||\, R_4 = \frac{2.2\ \text{k} \cdot 0.990\ \text{k}}{2.2\ \text{k} + 0.990\ \text{k}} = 0.6828\ \text{k}$$

$$R_2 \,||\, 0.6828\ \text{k} = \frac{5.7\ \text{k} \cdot 0.6828\ \text{k}}{5.7\ \text{k} + 0.6828\ \text{k}} = 0.6098\ \text{k}$$

$$R_{EQ} = R_1 \,||\, 0.6098\ \text{k} = \frac{4.1\ \text{k} \cdot 0.6098\ \text{k}}{4.1\ \text{k} + 0.6098\ \text{k}} = 0.5308\ \text{k}\Omega$$

3.

$$R_3 \,||\, R_4 = \frac{33\ \text{k} \cdot 33\ \text{k}}{33\ \text{k} + 33\ \text{k}} = 16.5\ \text{k}$$

$R_{EQ} = R_1 + R_2 + 16.5\ \text{k} = 13\ \text{k} + 27\ \text{k} + 16.5\ \text{k} = 56.5000\ \text{k}\Omega$

4.

$$R_4 + R_5 = 630\ \Omega\ + 630\ \Omega\ = 1{,}260\ \Omega$$

$$R_3 \,||\, 1{,}260\ \text{k} = \frac{470\ \Omega\ \cdot 1{,}260\ \Omega}{470\ \Omega\ + 1{,}260\ \Omega} = 342.3121\ \Omega$$

$$R_2 + 342.3121\ \Omega\ = 220\ \Omega\ + 342.3121\ \Omega\ = 562.3121\ \Omega$$

$$R_{EQ} = R_1 \,||\, 562.3121\ \Omega\ = 540\ \Omega\ \,||\, 562.3121\ \Omega$$

$$R_{EQ} = \frac{540\ \Omega\ \cdot 562.3121\ \Omega}{540\ \Omega\ + 562.3121\ \Omega} = 275.4651\ \Omega$$

5.

$$R_2 \,||\, R_3 = \frac{0.970\ \text{M} \cdot 0.860\ \text{M}}{0.970\ \text{M} + 0.880\ \text{M}} = 0.4558\ \text{M}$$

$$R_{EQ} = R_1 + 0.4558\ \text{M} + R_4 = 15\ \text{M} + 0.4558\ \text{M} + 11\ \text{M} = 26.4558\ \text{M}\Omega$$

6.

$$R_3 + R_4 = 6.1\ \text{k} + 1.3\ \text{k} = 7.4\ \text{k}$$

$$R_2 \,||\, 7.4\ \text{k} = 2.4\ \text{k} \,||\, 7.4\ \text{k} = \frac{2.4\ \text{k} \cdot 7.4\ \text{k}}{2.4\ \text{k} + 7.4\ \text{k}} = 1.8122\ \text{k}$$

$$R_{EQ} = R_1 \,||\, 1.8122\ \text{k} = 1.5\ \text{k} \,||\, 1.8122\ \text{k}$$

$$R_{EQ} = \frac{1.5\ \text{k} \cdot 1.8122\ \text{k}}{1.5\ \text{k} + 1.8122\ \text{k}} = 0.8207\ \text{k}\Omega$$

7.

$$R_3 + R_4 = 16\ \Omega\ + 22\ \Omega\ = 38\ \Omega$$

$$R_2 \,||\, 38\ \Omega\ = \frac{47\ \Omega\ \cdot 38\ \Omega}{47\ \Omega\ + 38\ \Omega} = 21.0118\ \Omega$$

$$R_{EQ} = R_1 \,||\, 21.0112\ \Omega\ = 15\ \Omega\ \,||\, 21.0118\ \Omega$$

$$R_{EQ} = \frac{15\ \Omega\ \cdot 21.0112\ \Omega}{15\ \Omega\ + 21.0112\ \Omega} = 8.7520\ \Omega$$

8.

$$R_3 \,||\, R_4 = 4.7\ \text{M} \,||\, 2.2\ \text{M} = \frac{4.7\ \text{M} \cdot 2.2\ \text{M}}{4.7\ \text{M} + 2.2\ \text{M}} = 1.4986\ \text{M}$$

$$R_{EQ} = R_1 + R_2 \,||\, 1.4986\ \text{M} = 0.950\ \text{M} + 1.2\ \text{M} \,||\, 1.4986\ \text{M}$$

$$R_{EQ} = \frac{2.15\ \text{M} \cdot 1.4986\ \text{M}}{2.15\ \text{M} + 1.4986\ \text{M}} = 0.8831\ \text{M}\Omega$$

Multi-Resistor Calculations 1

1.

$$R_{EQ} = 1\text{ k} + 2\text{ k} = 3\text{ k}$$

$$I = \frac{V_B}{R_{EQ}} = \frac{6\text{ V}}{3\text{ k}} = 2.0000\text{ mA}$$

$$V_1 = IR_1 = 2.0000\text{ mA} \cdot 1\text{ k} = 2.0000\text{ V}$$

2.

$$I = \frac{V_B}{R_2} = \frac{9\text{ V}}{8\text{ k}} = 1.1250\text{ mA}$$

3.

$$R_2 \,||\, R_3 = 10\text{ k} \,||\, 5\text{ k} = \frac{10\text{ k} \cdot 5\text{ k}}{10\text{ k} + 5\text{ k}} = 3.3333\text{ k}$$

$$R_{EQ} = R_1 + 3.3333\text{ k} = 5\text{ k} + 3.3333\text{ k} = 8.3333\text{ k}$$

$$I_1 = \frac{V_B}{R_{EQ}} = \frac{9\text{ V}}{8.3333\text{ k}} = 1.0800\text{ mA}$$

$$V_1 = I_1 R_1 = 1.0800\text{ mA} \cdot 5\text{ k} = 5.4000\text{ V}$$

$$V_3 = 9\text{ V} - V_1 = 9\text{ V} - 5.4000\text{ V} = 3.6000\text{ V}$$

$$P_{R3} = \frac{V_3^2}{R_3} = \frac{(3.6000\text{ V})^2}{5\text{ k}} = 2.5920\text{ mW}$$

4.

$$R_2 \,||\, R_3 = \frac{2\text{ k} \cdot 2\text{ k}}{2\text{ k} + 2\text{ k}} = 1\text{ k}$$

$$R_{EQ} = R_1 + 1\text{ k} + R_4 = 2\text{ k} + 1\text{ k} + 4\text{ k} = 7\text{ k}$$

$$I_1 = \frac{V_B}{R_{EQ}} = \frac{6\text{ V}}{7\text{ k}} = 0.8571\text{ mA}$$

$$V_4 = I_1 R_4 = 0.8571\text{ mA} \cdot 4\text{ k} = 3.4284\text{ V}$$

Multi-Resistor Calculations 2

1.

$$R_2 \parallel R_3 = \frac{2.2\text{ k}\cdot 4.5\text{ k}}{2.2\text{ k}+4.5\text{ k}} = 1.4776\text{ k}$$

$$R_{EQ} = R_1 + 1.4776\text{ k} = 1.5\text{ k} + 1.4776\text{ k} = 2.9776\text{ k}\Omega$$

$$I_1 = \frac{V_B}{R_{EQ}} = \frac{5\text{ V}}{2.9776\text{ k}} = 1.6792\text{ mA}$$

$$V_1 = I_1 R_1 = 1.6792\text{ mA}\cdot 1.5\text{ k} = 2.5188\text{ V}$$

$$V_2 = V_B - V_1 = 5\text{ V} - 2.5188\text{ V} = 2.4812\text{ V}$$

$$I_{R2} = \frac{V_2}{R_2} = \frac{2.4812\text{ V}}{2.2\text{ k}} = 1.1278\text{ mA}$$

2.

$$R_2 + R_3 + R_4 = 0.9\text{ k} + 1\text{ k} + 2.1\text{ k} = 4\text{ k}$$

$$R_{EQ} = 4.3\text{ k} \parallel 4\text{ k} = \frac{4.3\text{ k}\cdot 4\text{ k}}{4.3\text{ k}+4\text{ k}} = 2.0723\text{ k}\Omega$$

$$I_1 = \frac{V_B}{R_{EQ}} = \frac{9\text{ V}}{2.0723\text{ k}\Omega} = 4.3430\text{ mA}$$

$$I_{R1} = \frac{V_B}{R_1} = \frac{9\text{ V}}{4.3\text{ k}} = 2.0930\text{ mA}$$

$$I_{R2} = I_1 - I_{R1} = 4.3430\text{ mA} - 2.0930\text{ mA} = 2.2500\text{ mA}$$

$$V_3 = I_{R2} R_3 = 2.2500\text{ mA}\cdot 1.0\text{ k} = 2.2500\text{ V}$$

3.

$$R_3 + R_4 = 3.3\text{ k} + 4.7\text{ k} = 8\text{ k}$$

$$R_2 \parallel 8\text{ k} = \frac{2\text{ k}\cdot 8\text{ k}}{2\text{ k}+8\text{ k}} = 1.6\text{ k}$$

$$R_{EQ} = R_1 + 1.6\text{ k} = 0.5\text{ k} + 1.6\text{ k} = 2.1000\text{ k}\Omega$$

$$I_1 = \frac{V_B}{R_{EQ}} = \frac{12\text{ V}}{2.1000\text{ k}} = 5.7143\text{ mA}$$

$$V_1 = I_1 R_1 = 5.7143\text{ mA}\cdot 0.5\text{ k} = 2.8572\text{ V}$$

$$V_2 = V_B - V_1 = 12\text{ V} - 2.8572\text{ V} = 9.1428\text{ V}$$

$$I_{R2} = \frac{V_2}{R_2} = \frac{9.1428\text{ V}}{2\text{ k}} = 4.5714\text{ mA}$$

$$I_{R3} = I_1 - I_{R2} = 5.7143\text{ mA} - 4.5714\text{ mA} = 1.1429\text{ mA}$$

$$P_{R4} = I_{R3}^2 R_4 = (1.1429\text{ mA})^2 \cdot 4.7\text{ k} = 6.1392\text{ mWV}$$

4.

$$R_2 + R_3 = 4.7\ \text{M} + 1.5\ \text{M} = 6.2\ \text{M}$$

$$R_1 \,||\, 6.2\ \text{M} = \frac{1.5\ \text{M} \cdot 6.2\ \text{M}}{1.5\ \text{M} + 6.2\ \text{M}} = 1.2078\ \text{M}$$

$$R_{EQ} = 1.2078\ \text{M}\Omega$$

$$I_1 = \frac{V_B}{R_{EQ}} = \frac{6\ \text{V}}{1.2078\ \text{M}} = 4.9677\ \mu\text{A}$$

$$V_B = V_1 = 6\ \text{V}$$

$$V_1 = I_2 R_1$$

$$I_2 = \frac{V_1}{R_1} = \frac{6\ \text{V}}{1.5\ \text{M}\Omega} = 4.0000\ \mu\text{A}$$

$$I_1 = I_2 + I_3$$

$$I_3 = I_1 - I_2 = 4.9677\ \mu\text{A} - 4.0000\ \mu\text{A} = 0.9677\ \mu\text{A}$$

$$V_2 = I_3 R_2 = 0.9677\ \mu\text{A} \cdot 4.7\ \text{M}\Omega\ = 4.5482\ \text{V}$$

$$V_3 = I_3 R_3 = 0.9677\ \mu\text{A} \cdot 1.5\ \text{M}\Omega\ = 1.4516\ \text{V}$$

Multi-Resistor Calculations 3

1.

$$R_3 \,||\, R_4 = 3\ \text{k} \,||\, 3\ \text{k} = 1.5\ \text{k}$$

$$R_{EQ} = R_1 + R_2 + 1.5\ \text{k} = 2\ \text{k} + 2\ \text{k} + 1.5\ \text{k} = 5.5\ \text{k}\Omega$$

$$I_1 = \frac{V_B}{R_{EQ}} = \frac{5.5\ \text{V}}{5.5\ \text{k}\Omega} = 1.0000\ \text{mA}$$

$$V_1 = I_1 R_1 = 1.0000\ \text{mA} \cdot 2\ \text{k} = 2.0000\ \text{V}$$

$$V_2 = I_1 R_2 = 1.0000\ \text{mA} \cdot 2\ \text{k} = 2.0000\ \text{V}$$

$$V_3 = V_B - V_1 - V_2 = 5.5\ \text{V} - 2.0000\ \text{V} - 2.0000\ \text{V} = 1.5000\ \text{V}$$

$$I_{R3} = \frac{V_3}{R_3} = \frac{1.5\ \text{V}}{3.0\ \text{k}\Omega} = 0.5000\ \text{mA}$$

2.

$$R_2 \,||\, R_3 = 0.550\text{ k} \,||\, 0.470\text{ k} = \frac{0.550\text{ k} \cdot 0.470\text{ k}}{0.550\text{ k} + 0.470\text{ k}} = 0.2534\text{ k}$$

$$R_{EQ} = R_1 + 0.2534\text{ k} + R_4 = 5.5\text{ k} + 0.2534\text{ k} + 1.2\text{ k} = 6.9534\text{ k}\Omega$$

$$I_1 = \frac{V_B}{R_{EQ}} = \frac{13.7\text{ V}}{6.9534\text{ k}\Omega} = 1.9703\text{ mA}$$

$$V_1 = I_1 R_1 = 1.9703\text{ mA} \cdot 5.5\text{ k}\Omega = 10.8367\text{ V}$$

$$V_4 = I_1 R_4 = 1.9703\text{ mA} \cdot 1.2\text{ k}\Omega = 2.3644\text{ V}$$

$$V_2 = V_B - V_1 - V_4 = 13.7\text{ V} - 10.8367\text{ V} - 2.3644\text{ V} = 0.4989\text{ V}$$

$$I_2 = \frac{V_2}{R_2} = \frac{0.4989\text{ V}}{0.550\text{ k}} = 0.9071\text{ mA}$$

$$P_{R2} = V_2 I_2 = 0.4989\text{ V} \cdot 0.9071\text{ mA} = 0.4526\text{ mW}$$

3.

$$R_2 \,||\, R_3 + R_4 = 37\ \Omega \,||\, 45\ \Omega + 22\ \Omega = \frac{37\ \Omega \cdot 67\ \Omega}{37\ \Omega + 67\ \Omega} = 23.8365\ \Omega$$

$$R_{EQ} = R_1 + 23.8365\ \Omega = 62\ \Omega + 23.8365\ \Omega = 85.8365\ \Omega$$

$$I_1 = \frac{V_B}{R_{EQ}} = \frac{6.6\text{ V}}{85.8365\ \Omega} = 0.0769\text{ A}$$

$$V_1 = I_1 R_1 = 0.0769\text{ A} \cdot 62\ \Omega = 4.7678\text{ V}$$

$$V_2 = V_B - V_1 = 6.6\text{ V} - 4.7678\text{ V} = 1.8322\text{ V}$$

$$I_2 = \frac{V_2}{R_2} = \frac{1.8322\text{ V}}{37\ \Omega} = 0.0495\text{ A}$$

$$P_{R2} = V_2 I_2 = 1.8322\text{ V} \cdot 0.0495\text{ A} = 0.0907\text{ W}$$

$$I_3 = I_1 - I_2 = 0.0769\text{ A} - 0.0495\text{ A} = 0.0274\text{ A}$$

$$V_3 = I_3 R_3 = 0.0274\text{ A} \cdot 45\ \Omega = 1.2330\text{ V}$$

$$P_{R3} = V_3 I_3 = 1.2330\text{ V} \cdot 0.0274\text{ A} = 0.0338\text{ W}$$

4.

$$R_2 + R_3 \,||\, R_4 = 1.78\text{ M} \,||\, 2.2\text{ M} = \frac{1.78\text{ M} \cdot 2.2\text{ M}}{1.78\text{ M} + 2.2\text{ M}} = 0.9839\text{ M}$$

$$R_{EQ} = R_1 + 0.9839\text{ M} = 1.1\text{ M} + 0.9839\text{ M} = 2.0839\text{ M}\Omega$$

$$I_1 = \frac{V_B}{R_{EQ}} = \frac{11.6\text{ V}}{2.0839\text{ M}\Omega} = 5.5665\ \mu\text{A}$$

$$V_1 = I_1 R_1 = 5.5665\ \mu\text{A} \cdot 1.1\text{ M}\Omega = 6.1232\text{ V}$$

$$V_4 = V_B - V_1 = 11.6\text{ V} - 6.1232\text{ V} = 5.4768\text{ V}$$

$$I_3 = \frac{V_4}{R_4} = \frac{5.4768\text{ V}}{2.2\text{ M}\Omega} = 2.4895\ \mu\text{A}$$

$$I_2 = I_1 - I_3 = 5.5665\ \mu\text{A} - 2.4895\ \mu\text{A} = 3.0770\ \mu\text{A}$$

$$V_2 = I_2 R_2 = 3.0770\ \mu\text{A} \cdot 0.910\text{ M}\Omega = 2.8001\text{ V}$$

$$P_{R2} = V_2 I_2 = 2.8001\text{ V} \cdot 3.0770\ \mu\text{A} = 8.6159\ \mu\text{W}$$

Chapter 13

Mirror and Lens Calculations

1.

$f = -8.00 \text{ in}$

Negative focal length used for convex mirror.

$$h_{object} = 4.50 \text{ ft} \cdot \frac{12 \text{ in}}{1 \text{ ft}} = 54.0 \text{ in}$$

$$d_o = 25.0 \text{ ft} \cdot \frac{12 \text{ in}}{1 \text{ ft}} = 300.0 \text{ in}$$

$d_i = ?$

$M = ?$

$$\frac{1}{d_o} + \frac{1}{d_i} = \frac{1}{f}$$

$$\frac{d_i d_o}{d_o} + \frac{d_i d_o}{d_i} = \frac{d_i d_o}{f}$$

$$d_i + d_o = d_i \cdot \frac{d_o}{f}$$

$$d_i - d_i \cdot \frac{d_o}{f} = -d_o$$

$$d_i (1 - \frac{d_o}{f}) = -d_o$$

$$d_i = \frac{-d_o}{1 - \frac{d_o}{f}} = \frac{d_o}{\frac{d_o}{f} - 1} = \frac{300.0 \text{ in}}{\frac{300.0 \text{ in}}{-8.00 \text{ in}} - 1} = -8.219 \text{ in}$$

$$M = -\frac{d_i}{d_o} = -\frac{-8.219 \text{ in}}{300.0 \text{ in}} = 0.0274$$

$$M \cdot h_{object} = 0.0274 \cdot 54.0 \text{ in} = 1.48 \text{ in}$$

2.

$$f = 6.0\ \text{m} \cdot \frac{100\ \text{cm}}{1\ \text{m}} \cdot \frac{1\ \text{in}}{2.54\ \text{cm}} = 236\ \text{in}$$

$$h_{object} = 5\ \text{ft}\ 6\ \text{in} = 66\ \text{in}$$

$$d_o = 2.0\ \text{m} \cdot \frac{100\ \text{cm}}{1\ \text{m}} \cdot \frac{1\ \text{in}}{2.54\ \text{cm}} = 78.7\ \text{in}$$

$$d_i = ?$$

$$M = ?$$

$$\frac{1}{d_o} + \frac{1}{d_i} = \frac{1}{f}$$

$$\frac{d_i d_o}{d_o} + \frac{d_i d_o}{d_i} = \frac{d_i d_o}{f}$$

$$d_i + d_o = d_i \cdot \frac{d_o}{f}$$

$$d_i - d_i \cdot \frac{d_o}{f} = -d_o$$

$$d_i(1 - \frac{d_o}{f}) = -d_o$$

$$d_i = \frac{-d_o}{1 - \frac{d_o}{f}} = \frac{d_o}{\frac{d_o}{f} - 1} = \frac{78.7\ \text{in}}{\frac{78.7\ \text{in}}{236\ \text{in}} - 1} = -118.1\ \text{in}$$

$$M = -\frac{-118.1\ \text{in}}{78.7\ \text{in}} = 1.50$$

$$M \cdot h_{object} = 1.50 \cdot 66\ \text{in} = 99\ \text{in}$$

$$\frac{99\ \text{in}}{12} = 8\ \text{ft}\ 3\ \text{in}$$

3.

$$f = 250.0 \text{ mm}$$

$$d_0 = 3{,}880 \text{ mm}$$

$$d_i = ?$$

$$\frac{1}{d_o} + \frac{1}{d_i} = \frac{1}{f}$$

$$\frac{d_i d_o}{d_o} + \frac{d_i d_o}{d_i} = \frac{d_i d_o}{f}$$

$$d_i + d_o = d_i \cdot \frac{d_o}{f}$$

$$d_i - d_i \cdot \frac{d_o}{f} = -d_o$$

$$d_i (1 - \frac{d_o}{f}) = -d_o$$

$$d_i = \frac{-d_o}{1 - \frac{d_o}{f}} = \frac{d_o}{\frac{d_o}{f} - 1} = \frac{3{,}880 \text{ mm}}{\frac{3{,}880 \text{ mm}}{250 \text{ mm}} - 1} = 267 \text{ mm}$$

4.

$$h_{object} = 4.10 \text{ ft} \cdot \frac{12 \text{ in}}{1 \text{ ft}} = 49.2 \text{ in}$$

$$w_{object} = 2.75 \text{ ft} \cdot \frac{12 \text{ in}}{1 \text{ ft}} = 33 \text{ in}$$

$$d_i = 267 \text{ mm} \cdot \frac{1 \text{ cm}}{10 \text{ mm}} \cdot \frac{1 \text{ in}}{2.54 \text{ cm}} = 10.512 \text{ in}$$

$$d_0 = 3{,}880 \text{ mm} \cdot \frac{1 \text{ cm}}{10 \text{ mm}} \cdot \frac{1 \text{ in}}{2.54 \text{ cm}} = 152.76 \text{ in}$$

$$h_i(image) = ?$$

$$w_i = ?$$

$$M = -\frac{d_i}{d_o} = -\frac{10.51 \text{ in}}{152.76 \text{ in}} = -0.06881$$

$$h_i = M \cdot h_{object} = -0.06881 \cdot 49.2 \text{ in} = 3.39 \text{ in}$$

$$w_i = M \cdot w_{object} = -0.06881 \cdot 33 \text{ in} = 2.27 \text{ in}$$

5.

$$d_o = 8.75\text{ ft}\cdot\frac{12\text{ in}}{1\text{ ft}} = 105\text{ in}$$

$$d_i = 15.5\text{ in}$$

$$f = ?$$

$$\frac{1}{d_o}+\frac{1}{d_i}=\frac{1}{f}$$

$$f\left(\frac{1}{d_o}+\frac{1}{d_i}\right)=1$$

$$f=\frac{1}{\frac{1}{d_o}+\frac{1}{d_i}}=\frac{1}{\frac{1}{105\text{ in}}+\frac{1}{15.5\text{ in}}}=13.5\text{ in}$$

6.

$$f = 28.00\text{ mm}$$

$$d_i = 28.20\text{ mm}$$

$$d_o = ?$$

$$\frac{1}{d_o}+\frac{1}{d_i}=\frac{1}{f}$$

$$\frac{d_o d_i}{d_o}+\frac{d_o d_i}{d_i}=\frac{d_o d_i}{f}$$

$$d_i + d_o = \frac{d_o d_i}{f}$$

$$d_o - \frac{d_o d_i}{f} = -d_i$$

$$d_o\left(1-\frac{d_i}{f}\right) = -d_i$$

$$d_o = \frac{-d_i}{1-\frac{d_i}{f}} = \frac{d_i}{\frac{d_i}{f}-1} = \frac{28.20\text{ mm}}{\frac{28.20\text{ mm}}{28.00\text{ mm}}-1} = 3{,}948\text{ mm}$$

$$d_o = 3{,}948\text{ mm}\cdot\frac{1\text{ m}}{1000\text{ mm}}\cdot\frac{1\text{ ft}}{0.3048\text{ m}} = 12.95\text{ ft}$$

7.

$f = 12.00 \text{ in}$

$d_o = 3.00 \text{ in}$

$d_i = ?$

$M = ?$

$$\frac{1}{d_o} + \frac{1}{d_i} = \frac{1}{f}$$

$$\frac{d_i d_o}{d_o} + \frac{d_i d_o}{d_i} = \frac{d_i d_o}{f}$$

$$d_i + d_o = d_i \cdot \frac{d_o}{f}$$

$$d_i - d_i \cdot \frac{d_o}{f} = -d_o$$

$$d_i(1 - \frac{d_o}{f}) = -d_o$$

$$d_i = \frac{-d_o}{1 - \frac{d_o}{f}} = \frac{d_o}{\frac{d_o}{f} - 1} = \frac{3.00 \text{ in}}{\frac{3.00 \text{ in}}{12.00 \text{ in}} - 1} = -4.00 \text{ in}$$

$$M = -\frac{d_i}{d_o} = -\frac{-4.00 \text{ in}}{3.00 \text{ in}} = 1.33$$

8.

$$w_{o(object)} = 35.0 \text{ mm}$$

$$d_i = 82.5 \text{ ft} \cdot \frac{12 \text{ in}}{1 \text{ ft}} = 990 \text{ in}$$

$$d_o = 7.50 \text{ in}$$

$$f = ?$$

$$w_i = ?$$

$$\frac{1}{d_o} + \frac{1}{d_i} = \frac{1}{f}$$

$$\frac{f}{d_o} + \frac{f}{d_i} = 1$$

$$f(\frac{1}{d_o} + \frac{1}{d_i}) = 1$$

$$f = \frac{1}{\frac{1}{d_o} + \frac{1}{d_i}} = \frac{1}{\frac{1}{7.50 \text{ in}} + \frac{1}{990 \text{ in}}} = 7.44 \text{ in}$$

$$M = -\frac{d_i}{d_o} = -\frac{990 \text{ in}}{7.50 \text{ in}} = -132$$

$$w_i = M \cdot w_o = 132 \cdot 35.0 \text{ mm} = 4{,}620 \text{ mm} = 4.62 \text{ m}$$

9.

$$D_{i(initial)} = 0.88 \text{ mm}$$

$$f_1 = 125 \text{ mm}$$

$$f_2 = 650 \text{ mm}$$

$$D_{e(xpanded)} = ?$$

$$M = \frac{f_2}{f_1} = \frac{650 \text{ mm}}{125 \text{ mm}} = 5.2$$

$$D_e = M \cdot D_i = 5.2 \cdot 0.88 \text{ mm} = 4.6 \text{ mm}$$

10.

$$f = 125.0 \text{ mm}$$

$$h_i = 8.00 \text{ ft} \cdot \frac{0.3048 \text{ m}}{1 \text{ ft}} \cdot \frac{1000 \text{ mm}}{1 \text{ m}} = 2,438.4 \text{ mm}$$

$$d_i = 27.0 \text{ ft} \cdot \frac{0.3048 \text{ m}}{1 \text{ ft}} \cdot \frac{1000 \text{ mm}}{1 \text{ m}} = 8,229.6 \text{ mm}$$

$$h_o = ?$$

$$\frac{1}{d_o} + \frac{1}{d_i} = \frac{1}{f}$$

$$\frac{d_o d_i}{d_o} + \frac{d_o d_i}{d_i} = \frac{d_o d_i}{f}$$

$$d_i + d_o = \frac{d_o d_i}{f}$$

$$d_o - \frac{d_o d_i}{f} = -d_i$$

$$d_o(1 - \frac{d_i}{f}) = -d_i$$

$$d_o = \frac{-d_i}{1 - \frac{d_i}{f}} = \frac{d_i}{\frac{d_i}{f} - 1} = \frac{8,229.6 \text{ mm}}{\frac{8,229.6 \text{ mm}}{125.0 \text{ mm}} - 1} = 127 \text{ mm}$$

$$M = -\frac{d_i}{d_o} = \frac{8,229.6 \text{ mm}}{127 \text{ mm}} = 64.8$$

$$h_i = M \cdot h_o$$

$$h_o = \frac{h_i}{M} = \frac{2,438.4 \text{ mm}}{64.8} = 37.6 \text{ mm}$$

CPSIA information can be obtained
at www.ICGtesting.com
Printed in the USA
LVOW09s2307090617
537491LV00067B/804/P

9 780996 677158